World Food, Pest Losses, and the Environment

AAAS Selected Symposia Series

World Food, Pest Losses, and the Environment

Edited by David Pimentel

Routledge
Taylor & Francis Group
LONDON AND NEW YORK

First published 1978 by Westview Press

Published 2018 by Routledge
52 Vanderbilt Avenue, New York, NY 10017
2 Park Square, Milton Park, Abingdon, Oxon OX14 4RN

Routledge is an imprint of the Taylor & Francis Group, an informa business

Library of Congress Number: 77-90418

ISBN 13: 978-0-367-21399-2 (hbk)

About the Book

This book focuses on current food shortages and on the impact of pests in reducing world food supplies. At present, total worldwide food losses from pests are estimated to be about 45 percent. Preharvest losses alone, from insects, plant pathogens, and weeds, are estimated at about 30 percent. Additional postharvest losses from microorganisms, insects, and rodents range from about 10 to 20 percent. The contributors present data on the extent of these kinds of crop losses and analyze immediate and long-term pest management strategies. Emphasis is given to an evaluation of the effectiveness of integrated controls and the various new nonchemical pest controls used to reduce crop and livestock losses. The current worldwide environmental problems associated with both large-scale pesticide use and other pest control methods are also analyzed, including the impact that increased use of pesticides can be expected to have on the human environment. While the data included are technical, the presentation and analysis will be of interest to both the scientific community and the general public.

Introductory Note

Papers included in this book were presented at the Symposium on "World Food, Pest Losses and the Environment" held at the American Association for the Advancement of Science meeting in Denver, Colorado, February 1977. The objective of the Symposium was to examine the magnitude of food losses due to pests and to explore the impact of various pest control strategies on pests and the environment. In addition, we evaluated the impact of pests and control strategies on world food supplies. These analyses underscore the interdependencies of food production, food storage and pest control to meet the food needs of the growing world population.

We appreciate the assistance of Ms. Nancy Goodman in assembling and indexing and Ms. Barbara Kosoff in typing some parts of the book. We also thank Dr. Kathyrn Wolff and Ms. Arleen Rogan of AAAS for their help in seeing that the results of the symposium were published.

Contents

List of Figures

Chapter 4

Chapter 5

Chapter 6

Chapter 7

Chapter 8

List of Tables

Chapter 7

Chapter 8

Chapter 9

Foreword

The *AAAS Selected Symposia Series* was begun in 1977 to provide a means for more permanently recording and more widely disseminating some of the valuable material which is discussed at the AAAS Annual National Meetings. The volumes in this *Series* are based on symposia held at the Meetings which address topics of current and continuing significance, both within and among the sciences, and in the areas in which science and technology impact on public policy. The *Series* format is designed to provide for rapid dissemination of information, so the papers are not typeset but are reproduced directly from the camera copy submitted by the authors, without copy editing. The papers are reviewed and edited by the symposia organizers who then become the editors of the various volumes. Most papers published in this *Series* are original contributions which have not been previously published, although in some cases additional papers from other sources have been added by an editor to provide a more comprehensive view of a particular topic. Symposia may be reports of new research or reviews of established work, particularly work of an interdisciplinary nature, since the AAAS Annual Meeting typically embraces the full range of the sciences and their societal implications.

WILLIAM D. CAREY
Executive Officer
American Association for the Advancement of Science

About the Authors

David Pimentel, professor of entomology at Cornell University, has published over 160 research papers in the areas of entomology, ecology, pest management, and agricultural science. He is currently chairman of both the National Advisory Council of Environmental Education of HEW and the Board on Science and Technology for International Development of the National Academy of Sciences.

J. Lawrence Apple, associate director of research and professor of plant pathology and genetics in the School of Agriculture and Life Sciences at North Carolina State University, has written and lectured widely on crop protection programs and pest management, especially in developing nations. He is coeditor of Integrated Pest Management *(Plenum Press, 1976).*

Ralph A. Bram, principal staff officer of Vector Biology, Surveillance, and Control at the U.S. Department of Agriculture, has written two books and some forty articles and produced a film in the field of medical and veterinary entomology.

Donald J. Calvert is a crop protection specialist in the Department of Entomological Sciences at the University of California at Berkeley. He is currently involved in an AID project to improve pest and pesticide management practices of developing countries and is the editor of an international newsletter on pest management.

Roger O. Drummond is currently location and research leader in the U.S. Livestock Insects Laboratory at the U.S. Department of Agriculture. He has published over 125 papers in veterinary entomology and is an international authority on the biology and control of ticks and cattle grubs.

William R. Furtick, dean of the College of Tropical Agriculture at the University of Hawaii, has published over 100 articles on weed science, crop protection, and agricultural administration. He is the former director of the International Plant Protection Center at Oregon State University and coauthor of Weed Control Methods Manual *(Oregon State University Press, 1971) and* Oregon Weed Control Handbook *(Oregon State University Press, 1968).*

Hans Guggenheim, director of the Wunderman Foundation, has published widely in the fields of art history, anthropology, and world food problems. He has also written and directed films and conducted fieldwork in these areas.

Nels Konnerup is a livestock diseases specialist at the U.S. Agency for International Development. He has done work in foreign animal diseases and pest control.

Max Milner is a senior lecturer and the associate director of the International Nutrition Policy and Planning Program, Massachusetts Institute of Technology. He was a member of the Food Advisory Panel, Office of Technology Assessment, U.S. Congress (1974) and a participant in the World Food Congress in Rome (1974). He has published widely, recently coauthoring Protein Resources and Technology: Status and Research Needs *(Avi Publishing Co., 1977).*

H.A.B. Parpia is a senior officer with the Agricultural Services Division of the Food and Agriculture Organization of the United Nations (FAO) in Rome. He was formerly director of the Central Food Technological Research Institute in Mysore, India. A food technologist, he has studied and published widely on the problem of post-harvest food losses.

John R. Pedersen, instructor in the Department of Grain Science and Industry at Kansas State University, is currently working under an AID contract to supply technical assistance in grain storage, processing and marketing to USAID missions and their host governments worldwide. He is the author of a number of publications on insect pests in stored products, problems in grain quality preservation, and sanitation in food and feed plants.

Marcia Pimentel, lecturer in the Division of Nutritional Sciences at the College of Human Ecology, Cornell University, has published in the field of food science and nutrition. She is coauthor of Dimensions of Food *(Harper & Row, 1975).*

Nevin S. Scrimshaw is professor of human nutrition and head of the Department of Nutrition and Food Science at the

Massachusetts Institute of Technology. He has published over 350 articles on various aspects of human and animal nutrition, nutrition and infection, agricultural and food chemistry, and public health. Most recently, he coauthored Protein Resources and Technology: Status and Research Needs *(Avi Publishing Co., 1977).*

Ray F. Smith, professor of Entomology at the University of California, Berkeley, is also chairman of the FAO Panel of Experts on Integrated Pest Control, director of the UC/AID Pest Management Project, and past president of the Entomological Society of America. He is the author of more than 250 publications on integrated pest control, international crop protection, and related fields.

1

Dimensions of the World Food Problem and Losses to Pests

David Pimentel and Marcia Pimentel

Introduction

The objective of this book is to examine current food shortages and the role pests play in reducing world food supplies. These losses include destruction of harvestable food by pest insects, pathogens, weeds, rodents and birds, and postharvest losses to microorganisms, insects and rodents. In this study we also examine the immediate and long-term pest management strategies to reduce pest losses. Emphasis is given to the role of integrated control and various "bioenvironmental" (nonchemical) pest controls to reduce crop and livestock losses to pests. At the same time an analysis is made of current environmental problems associated with pesticide use and other pest controls.

To provide a perspective of the dimensions of the world food problem and losses to pests and the environment, in this chapter we examine the interdependencies of food, population, pests and environmental resources.

Patterns of Population Growth

Population

At no previous time in history have humans so dominated their environment by the sheer numbers of their species. This phenomenon, however, is a fairly recent event. For the first 990,000 years of the more than 1 million years that humans have inhabited the earth, the maximum world population numbered about 200,000 (about 1/4th the current population of Boston). For most of that period population growth was only about 0.01% per year (Figure 1). That comparatively slow growth rate accounts for why the world population took 990,000 years to reach about 200,000. During that long period of time people were hunter-gatherers

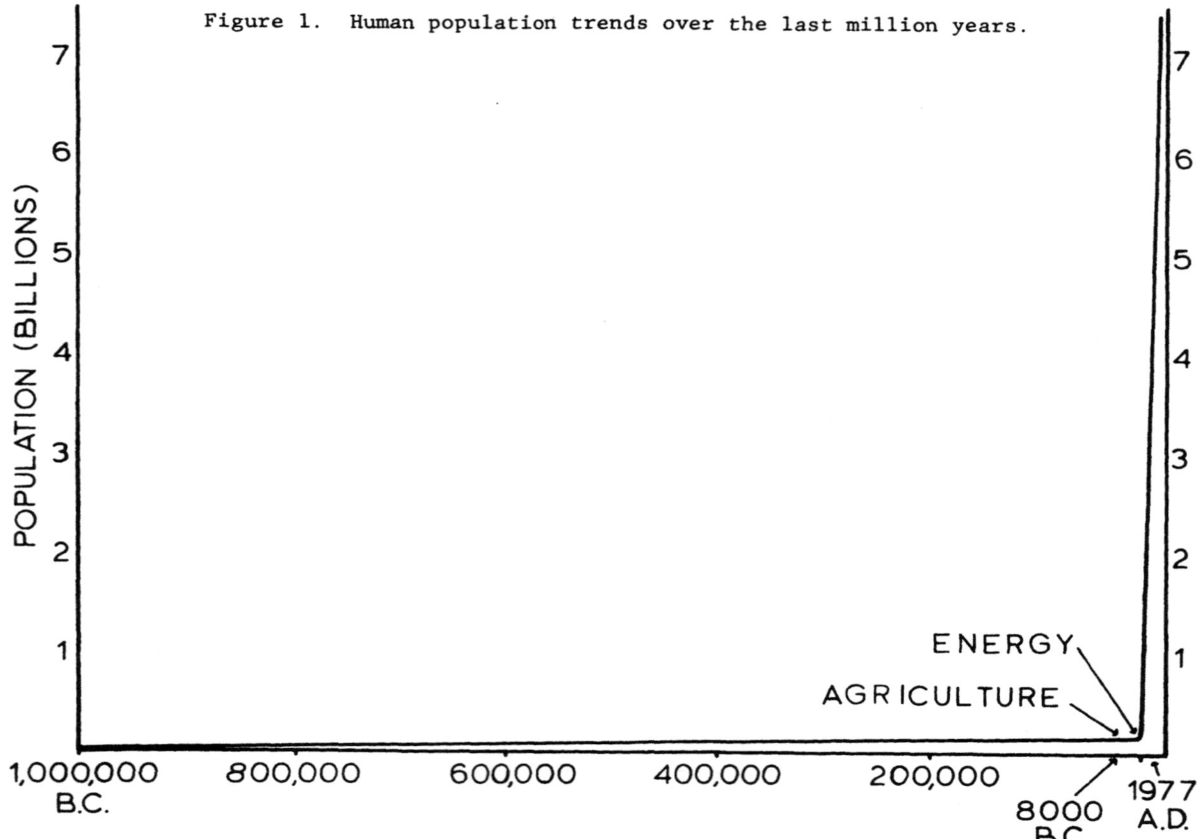

Figure 1. Human population trends over the last million years.

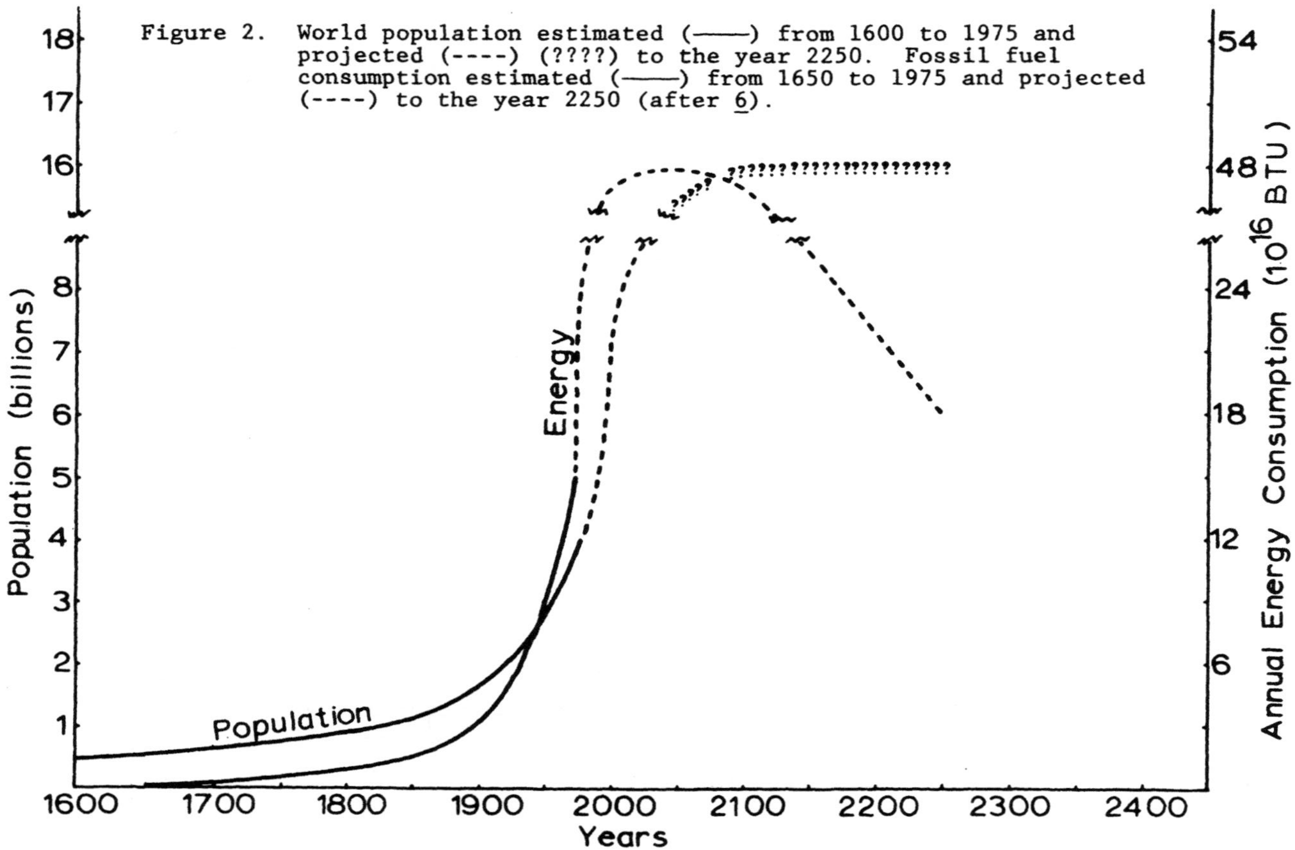

Figure 2. World population estimated (——) from 1600 to 1975 and projected (----) (????) to the year 2250. Fossil fuel consumption estimated (——) from 1650 to 1975 and projected (----) to the year 2250 (after 6).

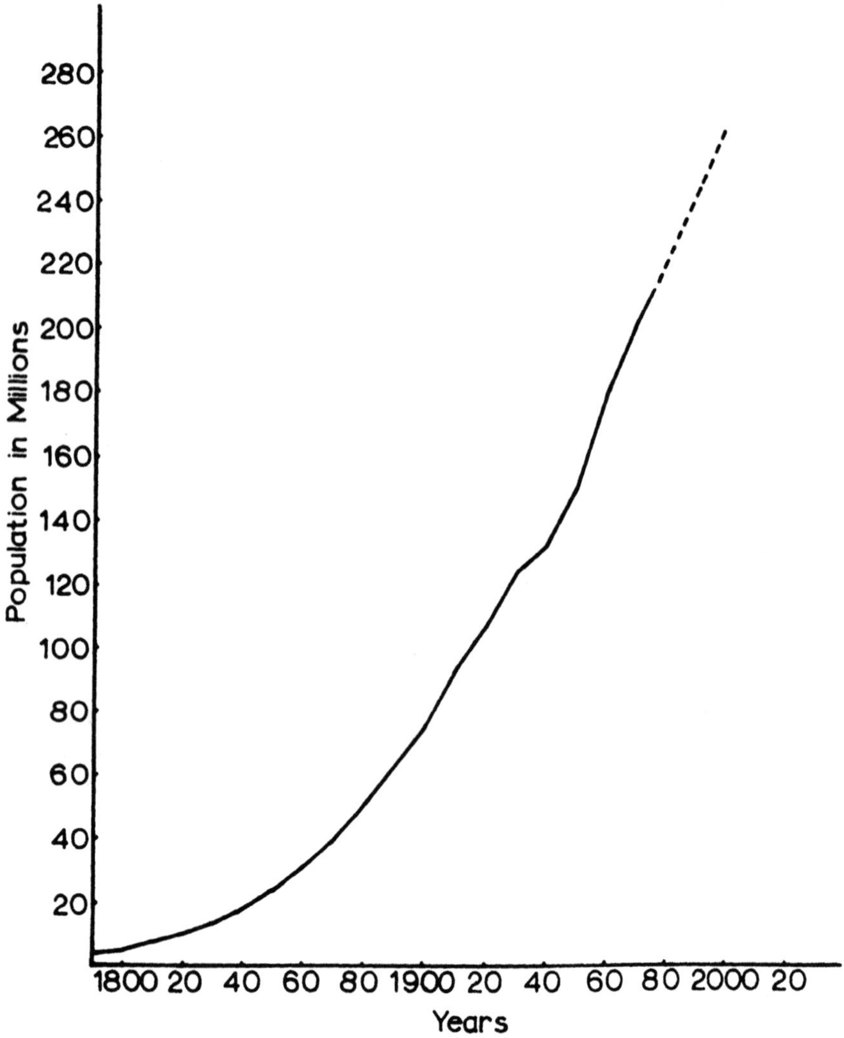

Figure 3. Population growth in the United States, actual (solid line) and projected (broken line) (after 38-40).

and depended exclusively on the natural ecosystem for their food.

Then about 10,000 years ago humans began to settle and cultivate food crops. With established agriculture came an increased and stable food supply. This contributed to the increased growth rate of the world population (Figure 1).

The next major increase in the population growth rate coincided with the discovery and use of stored fossil energy resources such as coal, oil and gas. Since that time, a mere 300 to 400 years ago, rapid population growth closely paralleled the increased use of fossil energy (Figures 1 and 2) because humans have been able to use energy to manipulate and manage their environment. In particular energy has been used to increase yields from food production and to improve public health. Improvements in both food and public health have been the prime contributors to rapid population growth in the world.

At present the world population stands at more than 4 billion, but what is more alarming is the high growth rate of nearly 2%. This is 200 times greater than the growth rate for the entire first 990,000 years of human existence on earth. This rate of increase contributes about 200,000 people to the world population daily.

Demographers project that the world population will reach 6 to 7 billion by the year 2000 and by 2100 will reach 10 to 16 billion (1). At present there appears to be no generally accepted way to limit this rate of growth (2).

The population in the United States is also projected to increase (Figure 3). The U.S. population is growing at about half the rate of the world population or about 0.9% per year (1). This rate is about 90 times the growth rate of the world population during the first 990,000-year period (Figures 1 and 3). During the next 25 years the U.S. population is projected to increase 24% from 212 million to 262 million. Although about 2/3rds of the increase will be due to immigration, 1/3 will be the result of the U.S. birth rate.

Contributing to the high growth rate of the world population is the young age-structure of many populations in many nations. For example, 52% of the population in Honduras is 15 years or younger (2). This means that a large proportion of the population is within childbearing age. This age structure has long-term implications for human population growth.

For example, in India where population density is already high, if all Indian families were to begin now to limit their family size to just two children, the population would continue to increase for the next 70 years (3). As a result the Indian population would increase from its present 600 million to more than 900 million. The same would be true of all countries with a high proportion of young in their populations and which restricted birth rates to just two replacement children per family.

Increased economic development is often cited as a solution to slowing birth rates. In the past, population growth rates have declined in some countries as incomes have risen (1). While this is encouraging and all would agree that economic development in all nations should be fostered, there is no evidence that future economic development alone will solve the world population problem.

The major reason why we cannot depend upon economic development only for population control is that humanity is already reaching the limits of the "carrying capacity" of environmental resources in many parts of the world (1). This is evidenced by the fact that populations in many parts of the world are already so large that they can no longer be fed without substantial food imports from other nations. Of the total of 149 nations in the world today, only 5 are net food exporters. The United States is, of course, the largest food exporter in the world (1).

Thus there is immediate need for us to take serious action to control our numbers before vital resources of our environment are insufficient to adequately support a world population. Indeed, it would be a tragedy for the credibility of humanity if a scarcity of resources limited human numbers.

Food is a vital resource that must supply the basic nutritional needs of the world population. Therefore, it is relevant to analyze the current status of food production and also to project how the world population of the year 2000 can be fed.

Current Food Supplies

Cereal grains are the primary source of food for humans. About half the protein and calories consumed by humans comes from the grains including wheat, rice, corn, millet, sorghum, rye and barley. About 90% of all the food for humans comes from these cereal grains plus cassava, sweet potato, potato, coconut, banana, common bean, soybean and peanut. In some developing nations the per capita consumption is about 400 pounds (182 kg) of grains or about 1 pound per day (1). This is just barely enough to support human life. In fact an estimated 1/2 billion humans are considered protein/calorie malnourished in the world today (4,5). In these areas the incidence of debilitating disease is also high.

The full dimensions of the current food problem are difficult to document. Reports indicate, however, that "famine in various developing nations, and death rates are reported rising in at least 12 and perhaps 20 nations, largely in Central Africa and Southern Asia" (1).

Certainly the disparity in caloric intakes is well documented. In a few countries like the United States, per capita caloric intake per day is about 3300 kcal and about 2/3rds of the protein consumed comes from animal sources. In contrast, the majority of the world population lives on about 2100 kcal per person per day and obtains most of its protein from plant proteins, especially grains (6). Thus it would seem that even at current production levels all humans are not receiving sufficient food. At the very least it is obvious that the demand for food in many nations is increasing at a rate equal to or greater than the increase in population.

Unfortunately world-wide crop yields per hectare have been declining rather than increasing (Figure 4). Even in the United States, with its usually high-yielding agriculture, setbacks have recently taken place. Heavy rains and flooding, droughts, crop pest problems and other problems have reduced our crop productivity. Other nations such as India, China and countries of Western Europe have experienced similar difficulties. Even without floods, droughts and pest problems, increasing crop yields has become a more difficult problem than in the past. This is because the prime resources for production—resources of arable land, water supplies and fossil energy—are becoming in shorter supply (7). Further, the long range effect of changing climatic conditions is surely a factor to be considered in the future (8).

Resources for Food Production

All agriculture and especially high-yielding agricultural production depends on basic environmental factors such as land, water, energy and climate.

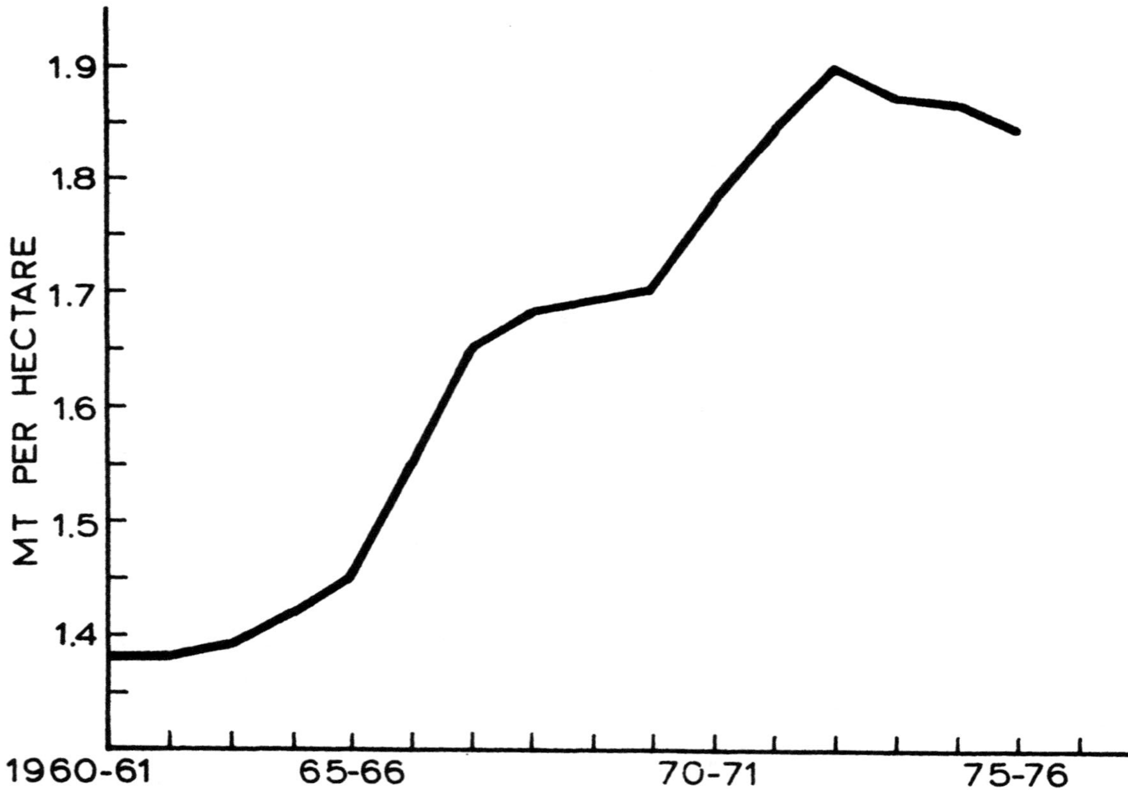

Figure 4. World grain yields per hectare (after 41). The equivalents are: a metric ton (MT = 2200 lb) and hectare (ha = 2.47 acres).

Land Resources

First consider that at present most of the land suitable for cultivation is now in use. Of the total 13.4 billion hectares of land area in the world, only about 11% or 1.5 billion hectares is considered suitable for cultivation (9). The remaining land is either too dry, too cold, too steep or lacks suitable soil structure for planting.

The situation in the United States provides some clarification of the magnitude of the problem. With about 160 million hectares planted to crops and a population of 212 million people, there is an average of about 0.77 hectares of cropland per capita. Since about 20% of our crops are exported, the estimated land actually used to provide food for our population is about 0.62 hectare (1.5 acres) per capita. Note that this amount of cropland is utilized to provide our high-calorie/high-animal protein diet and the production employs our energy-intensive techniques. Is there sufficient arable land to feed even 4 billion humans a protein-calorie diet similar to that of the United States employing U.S. agricultural technology? With about 1.5 billion hectares of arable land and a current world population of 4 billion, this averages to about 0.38 hectares (0.9 acres) of cropland per capita. Because this is about half the land used per capita to support the U.S. food consumption patterns, there is insufficient land available to feed the current world population a U.S. diet.

If indeed the world population increases to reach the projected 16 billion level, then world per capita arable land would be reduced from 0.38 hectares to only 0.1 hectares (less than 0.25 acres per capita). This estimate assumes increased population growth but no net reduction in agricultural land. But in fact with such a population expansion we logically expect a reduction in available land. For example, this has already happened in the United States, where from 1945 to 1970 over 29 million hectares have been lost to highways and urbanization, about half of which was valuable cropland (10,11).

Soil erosion seriously reduces the productivity of land. As a result an estimated 80 million hectares in the United States has been either totally ruined for crop production or has been so seriously eroded that the land is only marginally suitable for production (7,12,13). The rate of soil erosion per hectare of cropland in the United States is estimated at 27 metric tons annually (14,15). This relatively high rate of soil erosion has resulted in removing at least 1/3 of the topsoil of cropland in use today (16). In developing countries the rate at which land is lost due to soil erosion is estimated to be nearly twice as severe as it is in the United States (17).

The reduced productivity of U.S. cropland due to soil erosion has been offset by using increased quantities of fossil energy in the form of fertilizers and other inputs (15). An estimated 122 liters of fuel equivalents per hectare is being used to offset the soil erosion loss on U.S. cropland (7). If energy supplies decline, will this expenditure continue to be possible?

Furthermore, soil erosion also degrades our reservoirs, rivers and lakes by depositing annually about 1.4 billion metric tons in these water bodies (18). Soil sediments, the associated nutrients (nitrogen, phosphorus, potassium, etc.) and pesticide residues often have an adverse effect upon stream fauna and flora (7).

About 22% (3.0 billion hectares) of the world land area is pasture and range and is used to graze a livestock population about equal to the present human population (9). Some livestock are also pastured in the world's forests that occupy 30% of the land area (9). Livestock, particularly the polygastric ruminants, are important convertors of forage grasses and shrubs into food suitable for humans.

The potential cropland areas of the world might be doubled by irrigation and other massive environmental manipulations, but these manipulations require immense inputs of both water and energy. The possible impact of employing these strategies is discussed in the next section.

Water Resources

The availability of water is critical not only for agriculture but for industrial and home use. In the United States during the 20th century, the total "withdrawal" of water for all uses has been doubling at about 20-year intervals until at present about 1,400 billion liters per day are used. Although approximately 35% of this is used for irrigation, only 8% of U.S. agricultural land is irrigated.

By the year 2000, projected water needs will be double the current use figure (19). However, because water tables in some areas of the country are receding due to the mining of water, it is doubtful that this projected need will be realized. This may mean that increasing the amount of irrigated land in some areas will not be possible.

On a world-wide basis only about 12% of cultivated land is now irrigated (20). Suggestions have been made that the world's potential arable land might be doubled with increased irrigation and implementation of other significant alterations of the natural ecosystem (21). Unfortunately, irrigation and other similar environmental manipulations require enormous expenditures of energy. For example, a liter of water weighs 1 kg and about 12.2 million liters (12,200 metric tons) of water are needed to produce 5,000 kg of corn per hectare in the sub-tropics (22). The energy cost to pump this water from a depth of about 90 meters is about 20.6 million kcal (23). Based on this estimate (20.6 million kcal/ha), using irrigation to double the arable land from 1.5 to 3.0 billion hectares would require 3,090 billion liters of fuel per year. Further, if known usable petroleum reserves were used solely for this irrigation, then known reserves would last only 20 years! This estimate does not include the energy costs of supplying the irrigation machinery (an additional 13% in energy [24]) or the environmental costs such as salination of soil and other major problems associated with irrigation (25).

Not all forms of irrigation are as energy-demanding as sprinkler irrigation but all are energy intensive. No doubt in the future irrigation will continue to be used in some areas to make land more productive. The high energy costs of irrigation will, however, limit its use and certainly its use cannot be expected to double the arable land.

Energy Resources

The use of a large supply of inexpensive energy to manipulate and manage the environment is probably the most important factor responsible for the rapid growth of the world population. As noted earlier, the exponential increase of human numbers directly coincides with the use of fossil energy (Figure 2). In particular, energy has been expended to reduce human death rates by effective public health measures and to supply the food needed by a growing population.

Energy and food are in short supply today primarily for the same reason—growing human numbers. In fact energy use has been increasing faster than the world population. For example while the U.S. population doubled in the past 60 years, its energy consumption doubled in the past 20 years; while world population doubled in the past 30 years, world energy consumption doubled in the past 10 years.

To date our energy needs have been met by using up our nonrenewable resource, fossil energy with the result that known supplies are rapidly dwindling. Known world reserves of petroleum and natural gas are expected to be more than half depleted within the next 23 years (year 2000); and more than half of the coal reserves are estimated to be depleted about the year 2100 (26). Concurrently the world population is expected to increase nearly four times.

Agriculture, in its most productive form, depends upon large expenditures of many kinds of energy. For example, the production of a hectare of corn in the United States requires about 8.7 million kcal of fossil energy per hectare (about 860 liters of fuel equivalents) (27). More than half of this energy input is attributed directly to fuel and the production of nitrogen fertilizer. Substitution of fossil fuel-consuming machinery for labor has reduced the labor input to only 17 hours per hectare. This is much less than the 1,144 man-hours needed to produce corn primarily by hand labor (28). Although the fossil fuel input (8.7 million kcal) to produce a hectare of corn is only about 15% of the solar energy input (64 million kcal) (27), our concern focuses on fossil energy because this is a finite resource.

Energy input into U.S. agriculture accounts for nearly 5% of the national expenditure of fossil energy (29). An additional 5% is used for food processing and another 5% for distribution and preparation (6,30,31). Therefore, about 15% of U.S. fossil energy expenditure is attributed to the food system.

Converting this to gasoline equivalents, an estimated 1,251 liters (330 gallons) of gasoline equivalents of energy are expended in the U.S. food system to feed one person per year (29). When projected to a world population of 4 billion consuming a U.S. diet, the equivalent of 5,004 billion liters of fuel would be expended per year. Then the question is how long would known petroleum reserves last balanced against this kind of use. If 76% of the estimated 86,912 billion liters of petroleum reserves can be converted into fuels, this would give us a useable petroleum reserve of 66,053 billion liters (32). Then assuming petroleum were the only source of energy for food production, this known reserve would last a mere 13 years. Obviously the real outlook is not this grim, but critical problems do face us in our use of fossil fuel for agriculture and in the many other ways we depend on this finite resource. Further this assessment shows that we cannot rely solely on techniques of the past to increase food supplies.

Losses of Crops to Pests

On a world-wide basis, food losses to pests are high. At present, world crop losses to pests are estimated to be about 35% (33). These losses include destruction by insects, pathogens, weeds, mammals, and birds. Mammal and bird losses appear to be more severe in the tropics and sub-tropics than in the temperate region, but these losses are still low compared to the 3 major pest groups: insects, pathogens, and weeds.

Further, available evidence tends to suggest that "green revolution" technology has intensified losses to pests (34,35). One reason for this increased crop loss is the increased susceptibility of the new high-yield varieties to insects, pathogens, and weeds. In the past farmers most often selected seeds from individual plants that survived best under the native cultural conditions and produced the best yield for planting another season. These genotypes naturally harbored alleles that were resistant to insects and pathogens and that were competitive with weeds (35). Also new "green revolution" varieties are often planted uniformly over wide areas. This genetic uniformity provides an ideal ecological environment for pathogens to evolve and severely attack these genotypes (35,36). In some areas programs have been developed for multiple cropping in an effort to increase food supplies from limited land resources. The result of this continuous crop culture has been increased pest outbreaks.

Postharvest losses are estimated to range from 9% in the United States (37) to 20% in some of the developing nations especially in the tropics. The prime pests of harvested foods are microorganisms, insects and rodents.

When postharvest losses are added to preharvest losses, total food losses due to pests are estimated to be about 45%. Thus, the pest populations are consuming and/or destroying nearly 1/2 of the world's food supply. Surely this is a loss that we cannot afford as we face world food shortages and an ever increasing world population.

The status of available land, water supplies and use, possible changes in climate, the finiteness of fossil fuels and the magnitude of pest damage to food crops all place serious contraints on our ability to support projected human populations.

References

1. Committee on World Food, Health and Population, Population and Food (National Academy of Sciences, Washington, D.C., 1975).
2. National Academy of Sciences, Rapid Population Growth (Johns Hopkins Press, Baltimore, 1971), vols. 1 and 2.
3. K. Gulhati, Science 195, 1300 (1977).
4. The World Food Situation and Prospects to 1985, Foreign Agricultural Economic Report No. 98 Department of Agriculture, Washington, D.C., 1974).
5. United Nations, World Food Conference, Assessment of the World Food Situation (FAO, Rome, November 1974).
6. D. Pimentel, W. Dritschilo, J. Krummel, J. Kutzman, Science 190, 754 (1975).
7. D. Pimentel, E. C. Terhune, R. Dyson-Hudson, S. Rochereau, R. Samis, E. Smith, D. Denman, D. Reifschneider, M. Shepard, Science 194, 149 (1976).
8. R. A. Bryson, "The How and Why of Climatic Change," Paper presented at Food Editors Symposium, Pineapple Growers' Association of Hawaii, Honolulu (3 April 1977).
9. Production Yearbook 1972 (Food and Agriculture Organization, Rome, 1973).
10. Agriculture and the Environment, Economic Research Service No. 481 (Department of Agriculture, Washington, D.C., July 1971).
11. Our Land and Water Resources, Current and Prospective Supplies and Uses, Economic Research Service Miscellaneous Publication No. 1290 (Department of Agriculture, Washington, D.C., 1974).
12. U.S. National Resources Board, Soil Erosion, A Critical Problem in American Agriculture, Land Planning Committee, Supplementary report (Government Printing Office, Washington, D.C., 1935), p. 5.
13. H. H. Bennett, Soil Conservation (McGraw-Hill, New York, 1939).
14. C. H. Wadleigh, "Waste in relation to agriculture and forestry," U.S. Dept. Agric. Misc. Pub. No. 1065 (March 1968).
15. T. R. Hargrove, Iowa Agric. Home Econ. Exp. Stn. Spec. Rep. No. 69 (1972).
16. P. Handler, Ed., Biology and the Future of Man (Oxford Univ. Press, New York, 1970).
17. E. W. Ingraham, A Query into the Quarter Century (Wright-Ingraham Institute, Colorado Springs, Colo., (1975).

18. U.S. Department of Agriculture Research Program Development and Evaluation Staff, A National Program of Research for Environmental Quality—Pollution in Relation to Agriculture and Forestry, a report prepared by a joint task force of the Department of Agriculture and directors of Agricultural Experiment Stations (Department of Agriculture, Washington, D.C., 1968).
19. Water Policies for the Future (National Water Commission, Washington, D.C., 1973).
20. Production Yearbook 1969 (Food and Agriculture Organization, Rome, 1970).
21. C. E. Kellogg, in Alternatives for balancing World Food Production Needs, E. O. Heady, Ed. (Iowa State Univ. Press, Ames, 1967), pp. 98-111.
22. H. Addison, Land, Water and Food (Chapman and Hall, London, 1961).
23. E. T. Smerdon, in Sprinkler Irrigation Association 1974 Annual Technical Conference Proceedings, Denver, February (Sprinkler Irrigation Assoc., Silver Springs, Md., 1974), pp. 11-15.
24. D. Pimentel, Energy use in world food production, Environmental Biology, Report 74-1 (Cornell Univ., Ithaca, N.Y., 1974).
25. C. Clark, The Economics of Irrigation (Pergamon Press, London, 1967).
26. M. K. Hubbert, in The Environmental and Ecological Forum 1970-1971 (U.S. Atomic Energy Commission Office of Information Services, Oak Ridge, Tenn., 1972), pp. 1-50.
27. D. Pimentel, "Energy Use in Cereal Grain Production", to be presented at First International Conference on Energy Use Management, Tucson, Arizona (24-28 October 1977).
28. O. Lewis, Life in a Mexican Village: Tepoztlán restudied (Univ. Illinois Press, Urbana, 1951).
29. D. Pimentel, in Scienza and Tecnica 76 (Mondadori, Milan, Italy, 1976), pp. 251-266.
30. E. Hirst, Science 184, 134 (1974).
31. J. S. Steinhart and C. E. Steinhart, Science 184, 307 (1974).
32. H. Jiler, Commodity Yearbook (Commodity Research Bureau, Inc., New York, 1972).
33. H. H. Cramer, Pflanzenschutznachrichten 20(1), 1 (1967).
34. S. Pradhan, World Sci. News 8(3), 41 (1971).
35. I. N. Oka, personal communication.
36. O. H. Frankel, World Agr. 19, 9 (1971).

37. U.S. Department of Agriculture, Agricultural Research Service, Losses in Agriculture, Handbook No. 291 (Government Printing Office, Washington, D.C., 1965).
38. The World Almanac and Book of Facts, 1974 (Newspaper Enterprise Associates, New York, 1974).
39. World Population Data Sheet (Population Reference Bureau, Washington, D.C., 1975).
40. Statistical Abstract of the United States (Bureau of the Census, Washington, D.C., ed. 96, 1975), pp. 6-7.
41. L.R. Brown, The Politics and Responsibility of the North American Breadbasket, Worldwatch Paper 2 (Worldwatch Institute, Washington, D.C., 1975).

2

Insect Pest Losses and the Dimensions of the World Food Problem

Ray F. Smith and Donald J. Calvert

Introduction

In the past 25 years, there have been dozens of studies and investigations focusing on the problems of world food supply and population growth. These studies have been conducted by international organizations, universities, government agencies, individuals, and other private sector groups. They may be exemplified by the series of surveys on the world food situation made by the Food and Agriculture Organization of the United Nations (1,2,3) and the U.S. Department of Agriculture (4,5), by the special studies of world food prospects conducted by the President's Science Advisory Committee (6), Iowa State University (7), the University of California (8), and the USDA (9); by the special articles and reports on the world food problem appearing in Time (10), Science (11), and Scientific American (12); and by the publication of many books devoted to this subject (13).

The format of these panel and committee reports, journal articles or magazine stories follows a similar pattern - there is a historical review of the problem, an analysis of the factors affecting the supply and demand for food and, sometimes, a projection of the short- and long-term trend in food production, consumption, and trade. Then, depending on the assumptions made, the kinds of statistical information used, and the psychological bent of the authors, conclusions and opinions about the world's ability to feed its growing population are made, ranging from guarded optimism to grim predictions of impending catastrophe.

Various events that have occurred during this past quarter century could be used to support these differing viewpoints, but it is especially the events of the last decade, what with its occurrences of regional crop failures,

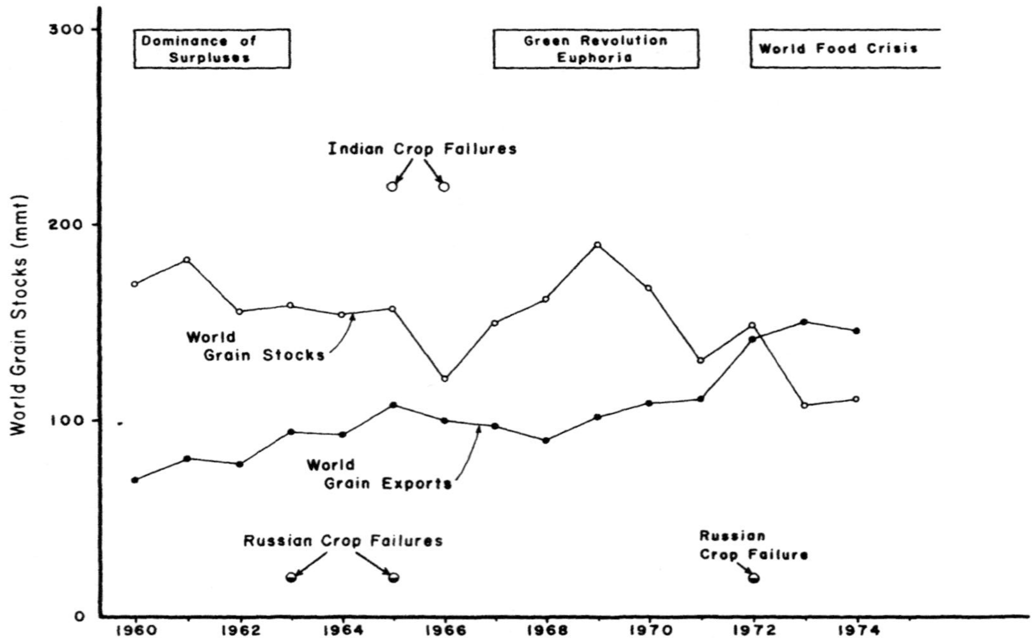

Figure 1 World grain stocks and exports from 1960–1974 (excluding rice and minor grains). [Modified from (37)]

the sharp reductions in world food reserves (Figure 1, Table 1) and the development of the energy crisis, that have heightened international concern over the problem of producing enough food for the world's expanding population to its present high level. Although world food stocks have been replenished somewhat once again following the excellent grain harvests in 1975 and 1976 (14), the long-term outlook remains discouraging. We must not forget that even larger food reserves were reduced rather quickly when bad weather caused a decline in world food production during 1972 and 1973.

World Food Production and Population Growth

The disheartening prospect for mankind is that periods of severe food shortage, which have been frequent enough in the past, appear likely to become even more frequent in the future because of the growing pressure on food supplies exerted by a continually expanding population. The attendant consequences of future shortages will become ever more serious unless the present rate of increase in population growth is sharply reduced or the amount of food produced greatly increased. However, producing enough food for the world's people is only a temporary stop-gap measure which merely postpones the time until a reduction in the rate of human population growth must also be achieved. In fact, the need to stabilize the human population at levels that can be sustained by the earth's resources, with proper consideration being given to the need for avoiding excessive destruction and contamination of the environment, remains the paramount problem facing mankind today.

Even though different conclusions on the world food problem may have been reached in the references cited, there has been general agreement that food production and population growth are closely interrelated problems greatly influenced in a complex way by social, economic and political factors. Because of these complex interactions and the magnitude of the problem, the eventual alleviation of hunger has been considered as a necessary long-term objective. Nevertheless, attainment of this objective will require immediate initiation of large technical assistance programs for the developing countries on a scale not previously attempted. Even if this effort is made, the consensus is that the situation cannot be expected to improve markedly in less than 5-10 years; therefore, inauguration of any proposed programs is urged as quickly as possible.

TABLE 1. World Food Stocks,[1]
1961-1976 (in million metric tons)

Year	Grain Food Stocks*	Reserves as Days of World Consumption
1960-1961	170	97
1961-1962	183	103
1962-1963	156	86
1963-1964	160	88
1964-1965	155	82
1965-1966	158	78
1966-1967	122	60
1967-1968	151	72
1968-1969	163	75
1969-1970	191	83
1970-1971	169	72
1971-1972	132	54
1972-1973	149	59
1973-1974	108	41
1974-1975	111	35
1975-1976+	100	31

*Excluding rice and minor grains.

+Preliminary estimates by USDA

[1]Data exclude countries for which adequate stock data are not available, e.g. USSR, the People's Republic of China (table modified after Walters, 1975).

Argument was recently advanced for the increased involvement of agricultural scientists from U.S. universities and experiment stations in the implementation of these technical assistance programs (15). Proponents feel that the increased participation of this large pool of experienced scientists would enable the developing countries to make significant progress in increasing their rate of food production while at the same time training local agricultural scientists. Again, increased funding would be necessary to bring about this augmented participation of U.S. scientists, but the prospects for a high social return on this investment clearly justify this expenditure.

It has been observed by many that the short-term task confronting biologists, administrators and government leaders is that of feeding the human population as a means of buying time until the stabilization of the population at a suitable level can be reached. However, some have warned (16) that in achieving this short-term goal we are contributing to further population increases which will result in a greater ultimate catastrophe in terms of human suffering if anticipated population control measures are not taken. Nevertheless, overall moral considerations require us to try to provide the time necessary for the larger goal to be achieved.

Until the population growth rate does stabilize, world food production will have to be accelerated considerably, especially in the developing countries, to keep up with the anticipated increases in population. This increase in production may be achieved in three ways: 1) expansion of the land area cultivated, 2) increase in yields obtained from the use of improved varieties, fertilizer, irrigation, etc., or 3) reduction in losses from pest attack. In the latter instance, tremendous increases in food supplies can be achieved rather rapidly simply through the adaptation and application of current technology. The following discussion concerns crop losses from pest attack with especial reference to insects. Other speakers will discuss the other pests and post-harvest losses.

Crop Losses in Rice

Insect pests can, and do, cause serious losses to many food crops in many different countries. These losses are known to be large in many cases and often are much greater than is generally appreciated. Cramer (17), in one of the very few global analyses of pest losses, undertook to assess and estimate annual world crop production losses caused by insect pests, plant diseases and weeds by reviewing the available literature up to 1966. Noting the incomplete-

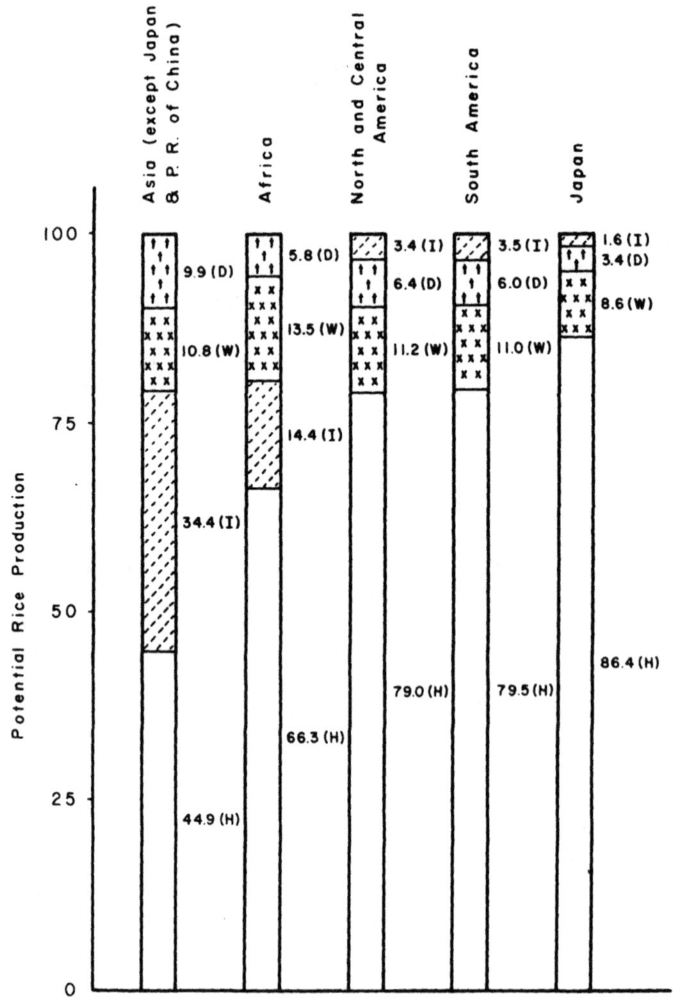

Figure 2. Estimated percentage potential rice production actually harvested (H) and percentage lost to insects (I), diseases (D), and weeds (W) in various parts of the world. [Based on (17), modified from (18)]

ness and inadequacy of much of the available data, he adopted a most conservative approach in making these estimates, but even so, the amount of damage he attributed to insect pests alone often ranged from 15-30% for most crops on a global basis.

To illustrate the impact of insect pests on world food production in more detail, the situation with rice provides an excellent example (18). Rice is the world's most important agricultural crop, providing a larger proportion of food for humans than any other single crop. It is the staple of existence for over 2 billion people.

In 1972, there were approximately 131.2 million hectares of land planted to rice from which over 295 million metric tons were harvested (19). There have been over 800 species of insects recognized as damaging rice in some way, although the majority of them do little relative damage (20). However, the few species which can cause serious damage are extremely important. Cramer (17) estimated that losses from insect pests of up to 35% regularly occurred in the Southeast Asia region. (Figure 2).

In tropical Asia, some 18-20 insect species are considered to be pests of major importance and regular occurrence. Those which are consistently considered to be the most destructive are stem borers, leafhoppers, planthoppers, rice bugs, gall midge, hispa beetle, leaf folder, armyworms, cutworms, and several species of maggots. Overall, the rice stem borers are probably the most serious group of insect pests. Twenty-one different species are known worldwide; of these, usually one to four species are important in any given area.

The enormity of the problem of stem borer damage is reflected in loss estimates from several rice-producing countries. At times, stem borers do so much damage in an area that total failure of the rice crop results. For example, in 1970 in parts of Pakistan fields could be harvested only for fodder or were turned over to grazing animals because there were essentially no filled heads of grain (21). In areas of India where two rice crops can be grown in a year, damage to the first crop may be so great that farmers become reluctant to make the effort to attempt a second one (22).

The rice leafhoppers and planthoppers, in addition to the direct damage they cause by sucking plant juices and injecting toxins into the plants, also indirectly cause

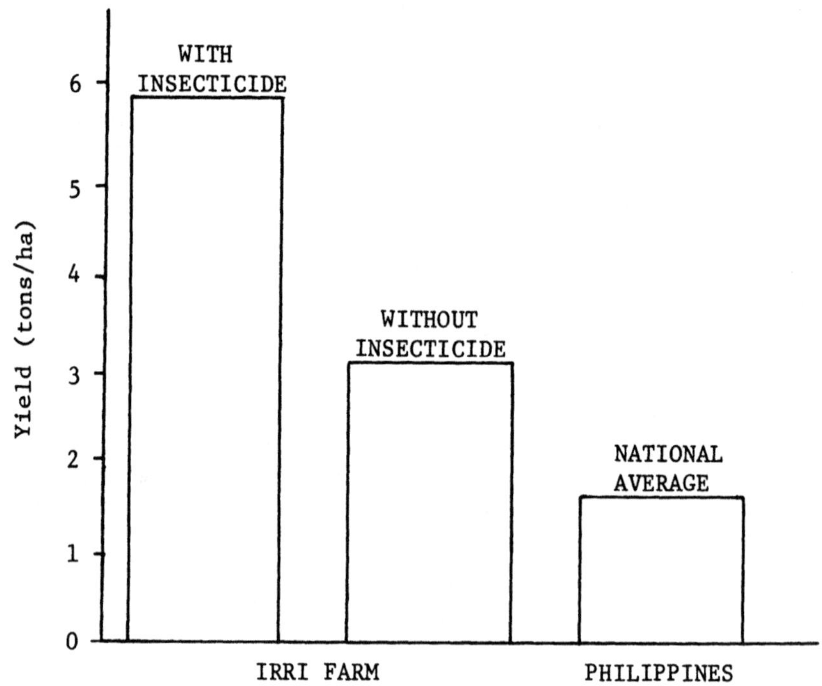

Figure 3. Rice yields from 1964-71 on treated and untreated experimental plots at IRRI compared with national farm average, Philippines [Data from (29)].

great damage to rice plants by transmitting diseases. An outbreak of rice leafhoppers in Bangladesh in 1956 was estimated to have caused 20-50% losses to crops at the heading stage and 50-80% on crops with unformed heads of grain (23). A planthopper infestation caused complete failure of an introduced rice variety over wide areas of two states in India in 1966 (24).

Losses of 10-40% from rice bugs have been reported from various countries and, in severe infestations, there may be complete loss of the crop (25). Grain damaged by these bugs may also affect the quality of the product when sold.

Armyworms and cutworms appear sporadically in enormous numbers. These pests often go unnoticed while they are small since most of the species usually feed at night and remain concealed in the daytime. After they become larger, they consume a great deal and may completely destroy a field in only one or two nights. The larvae commonly move from field to field in great numbers, eating all crops as they go. In 1966, an outbreak of armyworms devastated about half of the rice fields in the central and northern sectors of Ghana (26). An armyworm epidemic seriously affected 10,000 hectares in Malaysia in 1967 (27). Following a severe infestation of armyworms in an area in India in 1969, the yield from some experimental plots in rice fields was only 370 kg/ha, compared with yields of 1125 kg/ha in plots where the armyworms had been killed with insecticide (28).

Usually, a rice field is infested with more than one kind of insect pest. Thus, it is very difficult to determine the specific amount of loss caused by each pest, and, sometimes, very general loss estimates are made which apply to the insect pest complex as a whole. Be that as it may, valuable insight into the magnitude of losses caused by an insect pest complex in its entirety is provided by experiments in which certain plots are protected from insects while others are not. The overall extent of losses caused by insect pests has probably been demonstrated most effectively by researchers in the Philippines. An extensive series of experiments conducted on plantings of the International Rice Research Institute (IRRI) showed that plots of rice protected from insects yielded an average of 2.7 tons/ha more than plots that were not protected (29) (Figure 3).

The significance of crop protection in rice may be further emphasized by the recent observation (30) that if the land devoted to rice production in India were to yield as much as that which is presently obtained in Japan, then the amount of rice produced would be enough to meet the rice needs for almost twice the present human population in Southeast Asia. One of the reasons given for the great differences in yield between these two countries was the amount of losses from all pests, and insects in particular, sustained by the rice crop in India as contrasted with the amount lost in Japan. India, with inadequate plant protection, annually loses over 36% of its rice crop to insects, while Japan, utilizing "maximum" crop protection measures, suffers less than a 2% loss in yield to insects annually (17).

The representative loss data we have just presented for some of the more significant insect pests on rice can be multiplied many times over if other pests, such as nematodes, weeds, birds, rodents or diseases are also considered. Rice is in fact the crop that suffers the most losses from pests (46.4%), having lost nearly 280 million metric tons in 1974 (31). Many other crops regularly suffer losses of over 20%.

The relatively recent introduction of the new high-yielding grain varieties to developing countries is an extremely important agricultural advance in their struggle to attain self sufficiency in food production. However, cultivation of the new varieties is causing significant changes in their agricultural practices since these varieties must be provided with a number of modern inputs, such as fertilizer and good water control, before their true yield potential can be realized.

As more experience with the high-yielding varieties has been obtained, agriculturalists in these countries have become increasingly aware that pest problems are often more intense in the changed agricultural systems than in the traditional ones. Some serious epidemics of diseases and insect pests have occurred partly because the first varieties distributed had been bred primarily for their yield qualities with little attention to their relative resistance or susceptibility to various diseases and insect pests.

An additional reason why pest problems are of particular concern in plantings of the new varieties is that the high-yielding varieties have little genetic variability between individual plants, thus enhancing the vulnerability of the crop to large-scale pest attack. Newer varieties have replaced local varieties over large areas, but should they become susceptible to a disease or insect pest, devastating losses could occur over the entire planted area.

The southern corn leaf blight epidemic on maize in the United States, together with the tungro and brown planthopper epidemics on rice in Asia are recent examples of this increased vulnerability. Increasing the availability of germplasm to plant breeders so as to increase the genetic diversity in crops should assist in decreasing this vulnerability (31).

Some of the inputs that are essential to the enhancement of yield in the new improved varieties unavoidably create conditions that increase the probability that the crop may be seriously damaged by pests. For example, the high-yielding rice varieties are usually planted closer together than traditional varieties; they are shorter and have more tillers and under high fertilization produce a lush growth of foliage. This low, dense, succulent plant layer with high humidity is a favorable environment to the survival and development of many pests. With the creation of new micro-environments in rice fields, the composition of the pest complex may also change, and pests once considered minor may thrive to the extent that they cause serious losses. For example, the rice gall midge usually reaches epidemic proportions in traditional varieties only once every three to five years, but where these varieties have been replaced by the high-yielding rice, it is abundant every year (32). Leafhoppers and planthoppers have also increased greatly in importance following the introduction of the new varieties. On the other hand, the levels of a few diseases and insect pests have decreased somewhat in the new varieties.

The crop loss estimates given in our rice examples above illustrate the inadequacy of existing crop protection practices in preventing the ravages of plant pests in many parts of the world. It is clear that if these losses could be reduced food supplies would be increased without bringing new land and other limited resources into play. A recent NAS study (31) calculated that the increased supplies of food that would be made available if only 20% of the current pest losses in rice could be prevented would be sufficient to provide the foundation of an adequate diet for 177 million people. The report further stated that it was not unreasonable to think that improved pest control practices could eliminate 50% of current crop losses. This would mean an increase in food production of about 25% over current production; in other words, total world food production could reach about 82% of its potential instead of the 65% currently obtained.

Pest control programs to prevent crop losses from occurring have generally relied in the past on the unilateral use of a single control procedure. Great reliance on the use of pesticides was especially prevalent for a brief period following World War II during which these chemicals gave outstanding examples of control. However, in far too many instances, this control has been disrupted by the development of insect populations resistant to pesticides and by the occurrence of undesirable residue levels on food crops, impact on wildlife, the rapid resurgence of target pest populations following treatment, and outbreaks of unleashed secondary pests (33). Furthermore, in many instances, pesticides are not within the economic reach of peasant farmers.

Alternative Pest Control Strategies

An alternative approach to the management of pest populations, and one which holds great promise for reducing losses from pest attack, has been to stress the integration of a multiplicity of methods into a flexible control program. This approach, known as integrated pest management (IPM), may be described in simplest terms as a broad ecological attack combining several tactics for the economic control and management of all pest populations.

Integrated control practices rely on combinations of biological, chemical, and cultural control methods, as well as the use of resistant or tolerant plant varieties. Knowledge of the interrelationships between pests, crops and the environment is needed in order to devise the best combination of techniques which will give adequate control of the pest and produce minimum impact on nontarget organisms and the environment. Integrated pest management systems are thus broad, ecologically- based control systems that use and manipulate multiple plant protection tactics in an effective and coordinated way.

The leading example of the kind of research effort needed to develop successful integrated pest management systems is the coordinated project entitled "The Principles, Strategies, and Tactics of Pest Population Regulation and Control in Major Crop Ecosystems," supported by NSF, EPA, USDA and several state agricultural experiment stations. In this project, some 260 state and federal scientists joined together in a major effort to develop new, reoriented, expanded, and closely coordinated research efforts seeking practical alternatives to the extensive use of broad-spectrum chemicals for control of certain pest complexes. The effort involved a comprehensive national research program concentrating on five major agricultural crops and on pests of pine trees. The major objective was to develop systems of pest management that optimize cost/benefit ratios on a long-term basis for both the farmer and society. The program provided the basic biological information required for broad, ecologically- based control techniques of the pest insects on these crops. Coordinated with control of other pests, it developed methods of systematizing all relevant information necessary in pest control decision-making. The expertise and methods devised should be expanded and reinforced so this approach can encompass more crops and attack the full spectrum of crop pests.

In the opinion of Ennis et al. (34), the application of current pest control technologies could achieve a 30-50% reduction in crop losses and thus be of tremendous benefit in helping to increase world food supplies significantly. They advocated the further integration of this IPM system into the entire crop production process. This would necessarily involve the cooperative effort of scientists representing many disciplines working together with grower organizations, commodity groups and public agencies. Synchronization of this technology with other production inputs could result in the development of pest control programs which would be specific for the particular requirements of different agricultural areas and regions, i.e., utilizing different approaches in different localities depending on the pest complex, crop, season, etc. Accommodating the pest control system to the fundamental framework of the local society, taking into account the structure of land tenure, religious beliefs, pricing and marketing systems, educational institutions, public policies and local laws and customs is a necessary prerequisite for the successful implementation and operation of any pest control effort.

Over the past fifteen years, it has been clearly demonstrated that the integrated pest management philosophy is a sound, practical and economical approach to plant protection. Two recent National Academy of Sciences' studies both strongly recommended a change in emphasis in pest control toward the integrated pest management concept (31, 35). Quoting from the NAS Food and Nutrition Study "Research on the combination of pest control technologies and their integration into crop production systems offers the highest potential of any research in crop protection for alleviating world hunger and for improving the human environment. Such systems of integrated pest management are adaptable to all pests on all crops in every region of the globe" (31).

Systems of integrated pest management have demonstrated high impact on crop productivity and great adaptability in their application. The principles and practices of integrated pest management can be transferred and adapted to different conditions, e.g., from temperate to tropical conditions, even if they are first developed within the technology-intensive agriculture of the United States. It must be emphasized, however, that transferability is not automatic and cannot be made without careful adaptive research. In particular, institutional arrangements upon which integrated pest management systems depend are less transferrable than the biological principles involved. Research and extension personnel must be sensitive to the problems created in transferring technologies developed in the United States to non-industrialized tropical countries.

Agriculture often has little or no prestige in developing countries and in most, people who work with their hands are considered socially inferior. Students and their parents usually want him or her to go into medicine, the law, or even engineering, rather than go into agriculture. This form of social snobbery, which still exists to some extent in the United States, is commonplace in developing countries. Leaders in developing countries must participate in efforts to "legitimize" agriculture as an honorable profession before the best brains can apply themselves to solving their agricultural problems.

A major problem in developing countries is the lack of confidence small farmers have in governments and, more specifically, agricultural experiment stations and their representatives. Inflated claims of the value of agricultural research are common. It is true that new technology can often give two, three or even tenfold yield increases on experiment stations, but in the hands of small farmers only modest or no increases are obtained. The technology generated may be too expensive or sophisticated for a small farmer to use. Research scientists who go to developing countries are often confused and frustrated by the limited adoption of their recommended new practices and the slight yield increases obtained by farmers. Jennings (36) believes this problem is directly related to the multiple stresses of water, weeds, and pests characteristic of farmers' fields in the actual production areas, but relatively absent on experiment stations. Pest protection practices must be tested in farmers' fields and under a variety of stress conditions. Pests are undoubtedly a major reason for these paradoxical results, and a major reason why the small farmer in his decision making process often rejects information generated by "modern" agricultural science.

It is almost universally recognized that extension services in most developing countries lag far behind research activities. Duplication of extension activities by numerous government and private agencies, emphasis and servicing of the large (rich) rather than the small (poor) farmer, together with poorly planned, equipped and financed extension and teaching programs is normal. Frequently, the extension agents in developing countries have never lived or worked on a farm and usually come from a stratum of society in which people do not actually till the soil with their own hands. It is not difficult to understand why these agents have no confidence in their ability to help the farmer, and this opinion is fully shared by the farmer. If this situation is to be improved, radical changes are needed in the extension services of most developing countries and especially in the training of extension agents. For example, in some countries a poor farmer's son has no opportunity to ever receive university training in agriculture whereas many of the persons that are educated as extension agents have no background in agriculture at all and acquire little appreciation of the total complexity of farming operations during their training. An essential requirement for the establishment of an effective extension program is the training of culturally sensitive agents who can gain the confidence and respect of farmers.

In the past, the training given to students from developing countries in the United States too often proved inappropriate for increasing food production in their home countries. This disheartening situation resulted largely from the students' adoption of the U.S. scientific community's attitudes towards the prestige of agriculture, extension work, basic vs applied research, appropriate subjects for graduate research, administrative and funding agency criteria for promotions and funding research. When a student returned to his home country and was confronted with a lack of solid organizational or institutional support, the attitude he had adopted in the U.S. was far too often reflected in poor productivity and unhappiness. This situation could be immeasurably improved if these students were provided in-service training through participation in well-designed, problem-oriented research projects in their home country, another country in the region or at one of the regional or international agricultural research centers. U.S. institutions could play an important role in this process if they would make a stronger commitment to international agricultural assistance efforts and devote some of their resources to the development and strengthening of their capability to carry out this international technical assistance.

Research on crop pests is highly location specific. Many pest organisms have not yet spread to all of the crop production areas where they might be expected to attack crops and cause losses. Many noxious weeds, plant diseases, insect pests, and pestiferous birds and rodents have not yet moved across the oceans, deserts, mountains and other land and water barriers separating agricultural production regions. Such pests pose a constant threat to crop production in regions from which they have thus far been excluded whether by accident or design. It may not be known whether the pest would adapt to conditions prevailing in a new area. Serious pests in one production region are often unimportant in others. Also, it is often observed that minor pests in their native habitat become major ones when released free of their natural enemies in a new location.

Obviously, we need to know as much about new pests as possible before they are introduced, rather than after they have arrived and caused great losses. It is equally obvious that we cannot risk bringing a new pest into a new area for research purposes except for the most specialized laboratory research under the most rigorous quarantine conditions. Consequently, it is necessary that such research be carried out in areas where the pest is already present, preferably in the native habitat of the pest species. Frequently, but not always, the pests associated with a crop are present in their most diverse and potentially virulent or destructive form at the center of distribution of the crop species. This center is also often the best source of host plant resistance and of the natural enemies of the pest. It is thus a natural field laboratory for the study of pest control measures and to study the greatest numbers of pests of a crop in their fullest diversity. It is in such areas of diversity that research projects on pest management of major crops should be conducted. Such activity would provide the greatest access to the greatest number of pests of a crop and also the greatest access to natural control factors and host plant resistance with the least need for quarantine restrictions and other restraints on the full manipulation of the biological and ecological factors concerned.

Mechanisms for Technological Transfer

To assist in the alleviation of the world's food problem, there is a great need to have new mechanisms whereby pest control technology can be transferred in a rapid, effective and permanent manner to the developing world. This applies to the large body of technological knowledge already available and that which will be produced by continued research activities. A variety of mechanisms have been tested in the past and many have been found to be inadequate. In particular, the institution to institution relationship which has been characteristic of so much of U.S. overseas technical assistance in the past has a variety of deficiencies. At this point, when many traditional technical knowledge transfer-systems are being questioned or phased out, it is important to establish new ones which can achieve the critically important information-transfer. Perhaps the new arrangements should be a combination of several mechanisms.

Four mechanisms, each of which has its own useful characteristics, are suggested here as possibilities that can be explored and further developed for future implementation. Not listed in priority sequence, they are: a) an international plant protection center based in the U.S., b) overseas centers to carry out special functions in crop protection, c) a consortium of crop protection institutions in the U.S., and d) U.S. participation in the FAO/UNEP Cooperative Global Programme in integrated pest control. Each of these has its own rather independent role. The implementation of all of them in a coordinated way is realistic.

The International Plant Protection Center. The Center would provide a core of permanent research and extension staff concerned with crop protection problems on a global basis. It would supplement the work of a variety of existing organizations in a valuable way by directly stimulating the enthusiasm of scientists for the needed pest control research.

The United States at present does not have such a center. The U.S. Agency for International Development when a need occurs must borrow staff from the Department of Agriculture or recruit on an *ad hoc* basis from the universities. The U.S. Department of Agriculture is constrained (under most conditions) to work only on problems in the United States. The international regional agricultural research centers (IRRI, CIAT, CIMMYT, etc.) as presently developed cannot give adequate attention to the entire complex of pest problems, nor can they develop the integrated pest management strategies needed if pest control is to have a serious impact on crop yields on the time scale needed. The proposed U.S. Center should be concerned with all kinds of pests and have a strong integrated pest management philosophy.

Overseas Centers in Plant Protection. These are essential to any broad overall program of plant protection for the world's food supply and, as described above, are also critical to the adequate protection of U.S. agriculture. They could be autonomous institutes or linked to the International Plant Protection Center proposed above. It is critical that U.S. scientists become familiar with the biology and other characteristics of important crop pests in the places where they now occur so that they can respond rapidly and effectively when they invade new regions. There are other special overseas needs, for example, an overseas insect virus laboratory which can develop microbial control agents not available in the United States. In some cases, pests occur in the United States without their natural enemies or with only a few of them and more effective ones need to be sought overseas. There are a number of crops that are important to the world food supply which are not grown in the United States or, if they are grown, then they do not occur here with their most significant pests. These crops include cassava, rice, peanuts, coconuts, sorghum, millet, and cowpeas, among others. In the interest of an adequate supply of food for the world, the U.S. should also assist in the development of plant protection for these crops.

A U.S. Consortium of Crop Protection Institutes. The greatest concentration of talent and experience in crop protection now exists in the U.S. universities and agricultural experiment stations. No single international center in the U.S., or overseas, can ever expect to have such an array of qualified people. On many occasions, it will become critically important to tap this talent bank to obtain an individual scientist or teams of scientists with particular specialized knowledge. The development of a consortium arrangement, involving a group of the key U.S. institutions, would provide the mechanism by which this valuable talent bank could become available to work on international crop pest problems. The group of U.S. universities (Cornell, North Carolina, California, Florida, Miami, and Oregon State) working together in the UC/AID Pest Management Project have shown on a small scale the effectiveness of this arrangement.

FAO/UNEP Cooperative Global Programme for the Development and Application of Integrated Pest Control in Agriculture. FAO, in cooperation with UNEP, has established a Cooperative Global Programme for the development and application of integrated pest control in agriculture. The FAO Panel of Experts in Integrated Pest Control, which elaborated the Cooperative Global Programme, said in part in its report: "Man is today facing one of his most crucial food crises in history and the traditional plant protection input of pesticides is simultaneously critically limited by supplies. The integrated pest control strategy has been demonstrated to have the potential 1) to minimize environmental contamination, 2) to alleviate the problem resulting from the pesticide shortage and the increasing costs of chemical pest control, and 3) to increase the production of food and fibers. The Panel therefore recommends that immediate steps be taken to provide the resources to initiate as much as possible of the proposed Cooperative Global Programme for the development and application of integrated pest control. . . On this basis, the Panel recommends Regional Programmes for integrated pest control on cotton and rice as of the highest priority followed by a programme for maize and sorghum." It was agreed that the FAO Panel would serve as a formal advisory body for the Cooperative Global Programme.

The Cooperative Global Programme offers a rational mechanism for the introduction of the integrated pest control approach into the developing world where it can have positive impact in those areas where most of the world's food is produced. It also offers the means whereby sophisticated pest control technology of the developed countries can be adapted to the ecological and social conditions prevailing there. The United States should participate in a major way in the FAO/UNEP Cooperative Global Programme.

Summary

In summary, the development of sound plant protection systems for implementation in less developed countries could reduce pest losses as much as 50%. This reduction would conserve the yield gains expected from the introduction and use of the new high-yielding grain varieties and thus help to ameliorate the world food problem. A final solution to this problem, however, also would require improvements to be made in such inter-related aspects as economic growth, income distribution, energy use and population growth. The latter issue should receive the highest priority because of the overriding influence it has on the remaining factors. Because of the immense complexity of the problem and the interrelationship of the factors contributing to it, a satisfactory denouement will require a major commitment and marshalling of forces on the part of all the world's governments. Only through prompt and effective action can it be hoped that man's population density will be stabilized at an acceptable level as a result of his own initiative rather than having it imposed upon him by external forces.

References and Notes

1. FAO, World Food Survey (Washington, D.C., 5 July 1946).
2. ---, Second World Food Survey (Rome, November 1952).
3. ---, Third World Food Survey (Freedom from Hunger Campaign Basic Study 11, Rome, 1963).
4. USDA, Economic Research Service, The World Food Budget, 1962 and 1966 (Foreign Agricultural Economic Report 4, October 1961).
5. ---, Economic Research Service, The World Food Budget, 1970 (Foreign Agricultural Economic Report 19, October 1964).
6. Report of the President's Science Advisory Committee Panel on the World Food Supply (Government Printing Office, Washington, D.C., 1967). Three volumes.
7. World Food Production, Demand and Trade, Iowa State University Center for Agricultural and Rural Development, Ames, Iowa, 1973.
8. A Hungry World: The Challenge to Agriculture, University of California Food Task Force, General Report, Berkeley, California, July 1974.
9. USDA, Economic Research Service, The World Food Situation and Prospects to 1985 (Foreign Agricultural Economic Report 98, December 1974).
10. "The World Food Crisis." Special Section, Time (November 11, 1974) pp 68-83.
11. Science 188, 4188 (1975).
12. Scientific American 235, 3 (1976).
13. Representative examples are Josue de Castro, The Geography of Hunger (Little, Brown, Boston, 1952); M. K. Bennett, The World's Food (Harper, New York, 1954); Sir John Russell, World Population and World Food Supplies (Allen and Unwin, London, 1954); W. Paddock and P. Paddock, Famine 1975 (Little, Brown, Boston, 1947); and L. R. Brown and E. P. Eckholm, By Bread Alone (Praeger, New York, 1974).
14. Joe Western, National Observer, 25 December, 1976, p 1.
15. M. D. Whitaker and E. B. Wennergren, Science 194, 4263 (1976).
16. C. B. Huffaker, in Concepts of Pest Management, R. L. Rabb and F. E. Guthrie, eds. (North Carolina State University, Raleigh, 1970) pp 227-242.
17. H. H. Cramer, Plant Protection and World Crop Protection (Farbenjabriken Bayer AG, Leverkusen, West Germany, 1967).
18. B. Barr, C. S. Koehler and R. F. Smith, in Crop Losses Rice: Field Losses to Insects, Diseases, Weeds and Other Pests (UC/AID Pest Management and Related Environmental Protection Project Special Report, University of California, Berkeley, 1975). 64 pp.

19. FAO Production Yearbook 1972 (U.N. Food and Agriculture Organization, Rome, 1972).
20. D. H. Grist and F. J. A. W. Lever, Pests of Rice (Longmans, Green, London, 1969).
21. C. S. Koehler, in West Pakistan Rice Improvement Reports, Report No. RW23 (Ford Foundation, mimeographed, 1970).
22. P. Israel and T. P. Abraham, in The major insect pests of the rice plant (Proceedings of a symposium at the International Rice Research Institute, Los Banos, Philippines, Johns Hopkins Press, Baltimore, 1964) pp 265-275.
23. M. Z. Alam, in ibid., pp 643-655.
24. R. K. Patel, Int'l Rice Comm. Newsl. 20, 1 (1971), pp 24-25.
25. A. S. Srivastava and H. P. Saxena, in The major insect pests of the rice plant (Proceedings of a symposium at the International Rice Research Institute, Los Banos, Philippines, John Hopkins Press, Baltimore, 1964), pp 525-548.
26. M. Agyen-Sampong, paper given at West Africa Rice Development Association (WARDA) Seminar on Plant Protection for the Rice Crop, 21-27 May 1973, Monrovia, Liberia (Mimeographed).
27. J. R. Dunsmore, Int'l Rice Comm. Newsl. 19, 1 (1970) pp 29-35.
28. M. L. Purohit et al., Indian J. Agric. Sci. 41 (1971), pp 69-70.
29. M. D. Pathak and V. A. Dyck, PANS 19 (1973), pp 534-544.
30. M. J. Way, Bull. Entomol. Soc. Amer. 22, 2 (1976) pp 125-129.
31. National Academy of Sciences, Food and Nutrition Study, (National Academy of Sciences, Washington, D.C., 1977).
32. FAO, Report of the ad hoc Panel on the Rice Gall Midge in Asia and the Far East, 20-27 September, 1973, Bangkok, Thailand (FAO Regular Programme No. RAFE 15) 21 pp.
33. R. F. Smith and H. T. Reynolds, in The Careless Technology - Ecology and International Development, M. T. Farvar and J. B. Milton, eds. (National History Press, New York, 1971).
34. W. B. Ennis, Jr., W. M. Dowler and W. Klassen, Science 188, 4188 (1975) pp 593-598.
35. National Academy of Sciences, Pest Control: An Assessment of Present and Alternative Technologies, National Academy of Sciences, Washington, D.C., 1975). 5 volumes.

36. P. R. Jennings, 1974. Plant Breeding, the Green Revolution, and Food Production in Developing Countries. Donald F. Jones Memorial Lecture, Connecticut Agricultural Experiment Station.
37. H. Walters, Science 188, 4188 (1975), pp 524-530.

3

Impact of Plant Disease on World Food Production

J. Lawrence Apple

Man has lived with local and regional food shortages throughout history, but the inequality of food demand and food supply reached alarming proportions on a global basis in 1974. A combination of factors reduced world grain reserves to their lowest levels in two decades. These factors included burgeoning populations in the developing countries; shortages and high cost of pesticides, fertilizers, and fossil fuels; adverse weather conditions (1); the cumulative effects of a series of pest problems (2,3); and rising consumption of beef in the developed countries, thus increasing per capita demand for feed grains. This resulted in a decline in world food production on a per capita basis during the period 1970-1974 (4). Although this trend was reversed on a world-wide basis in 1976, many developing countries experienced either static food production or a decline on a per capita basis for the period 1970-1976 (5). Examples of such countries in the Western Hemisphere are Peru, Paraguay, Ecuador, Chile, Bolivia, Honduras, and Panama. Even Mexico has experienced an average decline in per capita food production during the 1970's. As a parallel to static or decreasing per capita food production, disposable income has increased in most countries on a per capita basis, thus widening the gap between effective demand for food and the available supply (6).

Although the actual magnitude of the world food shortage was and still is not known, famine has been reported in many developing countries (7). This imbalance in the world food/people equation has focused unprecedented attention on the need for increased agricultural production in both developed and developing nations. There are indications, however, that intensification of agricultural production also enhances the disease vulnerability of crop plant populations. But as the potential for crop damage due to diseases is probably increasing, the world can less afford these production losses because shortages in the world food supply are no longer buffered to the extent of previous years by surpluses in the developed nations.

World food production is reduced significantly each year by plant diseases, but unfortunately we have few reliable estimates of the magnitude of that loss. Most of this loss results from chronic, endemic diseases that do not kill individual plants but which produce a debilitating effect with the result of production loss. Some of these effects go unnoticed by farmers and gardeners because the pathogens are either systemic (such as viruses, mycoplasma, some fungi, and bacteria), or they attack the below-ground plant parts. But the history of plant pathology records many disease epidemics that have caused great production loss and famine. The late blight disease epidemic of the Irish potato in Europe in 1845 is one of the classic examples that contributed to the death of a million people in Ireland and the emigration of another million. More recently we recall the Southern Corn Leaf Blight epidemic of 1970 in the United States. That event caused a greater production loss on a single crop in one year than any similar event in the history of agriculture. The value of the lost production was $1 billion based on 1970 prices (3). It is fortunate that an epidemic of this magnitude did not occur in a developing country because it could have caused human suffering comparable to the potato famine of Europe in 1845, but even so it was not without its effects. It resulted in a critical reduction of our grain reserves, higher feed-grain and food prices, and set the stage for a world-wide food shortage that worsened during the succeeding years. Such dramatic and devastating disease events capture public and official attention. Their recurrence must be avoided if possible, but the losses produced by the less dramatic endemic diseases that are under reasonably good control, either by man-directed or natural forces, must be quantified and recognized as a major factor in reduced world food production. Methods for disease-loss assessment on a regional or national scale are inadequate, but there is renewed interest in developing improved methods through mathematical modeling, remote sensing and other techniques. Hopefully their practical utilization will be possible within a few years. Reliable disease loss estimates are requisite to the development of rational and economical crop protection systems. Without accurate disease-loss appraisals, we lack an objective basis for establishing research priorities and for establishing cost/benefit relationships that are necessary in applying disease management tactics at the farm level. With full recognition of the fact that the reliability of available disease-loss data is uncertain, we shall examine the best subjective assessments available for the United States and the world.

Food Losses Due to Diseases in the United States

The latest comprehensive assessment of production losses due to plant pathogens in the United States was published by the U. S. Department of Agriculture in 1965 which reflected losses during the period 1951-1960 (8). The estimated loss was approximately $4 billion annually due to diseases (including those caused by nematodes). This potential production loss occurred even though an estimated $131 million was spent annually during that period to control plant diseases! Major crops such as wheat, corn, and soybeans sustained losses in the range of 14%. Most vegetable and fruit crops reflected losses even higher -- in the range of 20% and above. As pointed out earlier, we lack accurate and practical techniques for assessing losses due to diseases on a regional or national basis. But granting that the accuracy of the above estimates may be questioned, they represent conservative judgments by experts in the various crop protection disciplines, and by any standard they are unacceptably high.

The Council on Environmental Quality reported (9) that crop losses in the United States due to plant diseases have increased since 1940, both in gross and percentage of crop value. Pimentel (10) suggested a slight reduction in disease losses between 1960 and 1974 from 12.2% to 12.0% of potential production; however, losses due to plant diseases either remain at a constant percentage of the crop or are actually increasing in the United States even though we have developed the most advanced crop protection technology in the world and a well developed information delivery system. The incongruity of this situation suggests either poor application of disease control methodology <u>or</u> increased vulnerability of modern crop production systems to disease attack <u>or</u> both. The implications of this situation will be explored later in this chapter.

World Food Losses Due to Diseases

The only compilation of disease losses on a global basis is that of Cramer published in 1967 (11). He also pointed out that the accuracy of available data are uncertain but he indicated that the data from the United States was probably more accurate than that from any other country. Cramer regarded his estimates of loss as conservative because he used the lower of available estimates and because he averaged over many years which leveled out the high losses caused in some years by epidemics. Cramer's data for major crops are summarized in Table 1. The average disease losses by crops for the world are generally in the same range as those for the United States <u>or even less</u> which attests to the relative ineffectiveness of our disease management technology or to the extreme vulnerability to disease of intensified agricultural production systems.

Table 1. Losses from diseases in world's major crops*

Crop	% Loss	Production-MMT-1974 Actual	Potential	Loss
Paddy Rice	8.9	323	354	31
Wheat	9.1	360	396	36
Maize	9.4	293	323	30
Millet/Sorgh.	10.6	93	104	11
Soybeans	11.1	57	64	7
Peanuts	11.5	18	20	2
Potatoes	21.8	294	376	82
Cassava	16.6	104	125	21
Sweet Potatoes	5.0	134	141	7
Tomatoes	11.6	36	41	5
Bananas	23.0	36	47	11

*The % loss data from Cramer (11) and the production data from FAO (12).

Having taken stock of what we judge to be the present state of disease losses in world agriculture, I wish to examine disease vulnerability of our modern agricultural production systems which we shall term agroecosystems. What is our perspective of plant pathogens in the agroecosystem?

Perspective of Plant Pathogens in the Agroecosystem

Modern plant disease control is based strongly on the use of chemicals, selected cultural practices, and host resistance. In utilizing these methods, ecological principles have been applied only sparingly in either interpreting or predicting behavioral patterns of hosts or pathogens. The lack of ecological perspective in agricultural research, especially crop protection research, has given rise to many false impressions and expectations.

Our perspective of agroecosystems and our research programs seldom reflect the fact that agricultural crops were selected by man from natural plant communities, that they have been changed in most instances to suit better his food and fiber needs (to such an extent, in fact, that few of them can now survive in natural plant communities), and that he has propagated them at epidemic population levels in very unnatural ecosystems. Agriculturists have learned over the centuries, principally by empirical processes, that inter-specific competition and low-yielding phenotypes of the cultivated species should be eliminated for maximum yields. These discoveries, augmented by the results of experimental science, have led to the present-day monocultures. They are a manifestation of the ecological principle that species simplicity rather than diversity is the most highly productive state for an ecosystem (13).

According to Preston (14), stability in the ecological world is not a static equilibrium but a fluctuating or dynamic one. Stability derives from the ability to bounce back. Preston considered an ecosystem stable during periods when no species became extinct and when none reached plague proportions (thus destroying the niches of other species). The agroecosystem must be considered unstable because the crop species is definitely in "plague" proportions, and it has little or no natural capacity for bouncing back except through the efforts of man. It is clearly a system that must be carefully managed if it is to sustain high productivity. It is not a system characterized by "biological balance" but by "biological imbalance". Thus, man's role is not to maintain biological balance of the agroecosystem (in ecological terms) but to maintain the dynamic state of imbalance that maximizes production of his crops.

Lack of understanding or appreciation for this concept of the agroecosystem is basic to a false concept of the status of plant diseases in agroecosystems. A false concept of both the state of the agroecosystem and of the status of plant diseases in that system is illustrated by such statements as: "The occurrence of a plant disease thus indicated that some aspect of the biological balance is not in equilibrium ..." (15).

It is within this conceptual milieu of the agroecosystem that we often attribute epidemics to "mistakes of the past" that have upset the biological balance. Such expressions convey the false impression that diseases are not expected to occur on cultivated crops and that epidemics result only when agroecosystems are mismanaged in a way that upsets the biological balance. The popular press has exploited these false perspectives by criticizing modern agricultural technology for creating ecological upsets that threaten man's food supply (16). An agroecosystem is not in a state of ecological or dynamic balance because it has many unfilled niches that would be invaded by competing organisms in the absence of intensive management efforts by man. The result would be a reduction of the dominance by the cultivated crop and reduced productivity. We must recognize and utilize these ecological principles as we seek to achieve even higher levels of agroecosystem productivity, especially as we seek to transform the traditional agricultural production systems found in many developing nations into more modern and highly productive ones.

The Challenge

An immediate challenge of the United States and the world is to optimize agricultural productivity per unit of land, water, fertilizer, energy and time. It has been shown experimentally that many of the practices that are used to increase crop productivity may also make the growing crop more susceptible to damaging attacks by plant pathogens. Such practices include: (a) high rates of fertilization; (b) high plant population density; (c) reduced genetic diversity within the plant population; (d) temporal and spatial monocultures; (e) improved water management; (f) changed tillage practices (e.g., minimum or no-till practices); and (g) negative or antagonistic interactions of crop protection practices. I wish to elaborate on some of these relationships to demonstrate the relatively higher disease vulnerability potential of the modern agroecosystem as compared to less-intensive, traditional systems found in many developing countries in the tropical regions.

The epidemiology of plant diseases follows the same general principles as diseases of man and other animals. Disease intensity in a population of plants is directly related to the density and other biological and physical characteristics of the suscept population over time. That is to say that the potential for damaging attack by a plant pathogen is related to the density of a susceptible plant population in a given region and in a given field, to the time the susceptible population is sustained in the same region, and to the biological and physical characteristics of the plant population and its surrounding environment. Some of these relationships are demonstrated by the data of Waggoner (17) in Tables 2-5. These data represent a foliar disease of crop plants caused by a fungus that produces and is disseminated by air borne spores. The data of Table 2 demonstrate the relationship between distance and the number of spores (inoculum) that would be transported from an inoculum source to an adjacent (target) field. When the fields are separated by 0.1 km, there is a 99.9% probability that plants in an adjacent field will become diseased (mean deposition of 280 "effective" spores on a 0.1 ha area) whereas separation of fields by 1 km gives a 25% probability that an adjacent field would not become diseased (see probabilities, Table 5). Larger fields also enhance the

Table 2. Relation of Distance of Source and Number of Spores Falling in Target Field*

Distance (km)	0.1	1.0
Thousands of Spore Lesions	1.0	1.0
Area of Target Field (ha)	0.1	0.1
Mean Deposition of Spores in Target Field	280.0	1.4

*Data from Waggoner (17)

chances of infection from a nearby inoculum source (Table 3). These data indicate a linear relationship between the chance for infection and size of field. There is also a linear relationship between the number of spores produced in the inoculum

Table 3. Relation of Size of Field to Number of Spores Falling in Target Field*

Distance (km)	1.0	1.0
Thousands of Spore Lesions	10.0	10.0
Area of Target Field (ha)	0.1	0.22
Mean Deposition of Spores	14.0	31.0

*Data from Waggoner (17)

source and the number of spores that would fall on an adjacent (target) field (Table 4).

In summary, Waggoner's data indicate that the concentration of spores drops away rapidly with distance from the inoculum source, that the danger from foreign infection increases with the spore concentration in a neighbor's field and with the proportion of the area occupied by fields of susceptible plants, and that increasing either the size of fields or the concentration of crops in an area increases the disease hazard.

Table 4. Relation of Concentration of Source and Number of Spores Falling in Target Field*

Distance (km)	1.0	1.0
Thousands of Spore Lesions	1.0	10.0
Area of Target Field (ha)	0.1	0.1
Mean Deposition of Spores	1.4	14.0

*Data from Waggoner (17)

These epidemiological relationships suggest that the traditional agricultural production system would be less subject to disease epidemics than modern agroecosystems. Traditional agriculture is in a state of ecological equilibrium under conditions of low fertility, poor moisture control, genetically heterogeneous mixtures of plants of low yield potential but

Table 5. Probabilities of (j) Spores Falling on the Field*

Mean Deposition of Spores (number)	280.0	31.0	14.0	1.4
%P(0)	0.1	4.0	0.1	25.0
%P(10)	0.1	0.1	6.6	0.1
%P(20-40)	0.1	94.0	7.6	0.1

*Data from Waggoner (17)

with high plasticity to adversity. These traditional varietal mixtures have been selected over long periods of time and are well adapted to these cultural conditions. They represent a compromise between yielding ability and genetic fitness to growing conditions. Actually crop species are often intermixed on the same ground. The fields are often small and scattered. Field sites may be shifted each 1-3 years as the inherent fertility becomes exhausted, especially in the humid tropical regions. All of these characteristics provide greater protection of the crop plants against disease epidemics. Plant diseases are found in these small, scattered fields of mixed plant types, but they are seldom damaged to the point of producing no harvest.

International agricultural development has been accelerated within the past 10 years by the production package of the "green revolution". This production system is based on high-yielding varieties along with improved tillage, water management, crop protection, and fertilization. The changed ecology coincident with the introduction of the green revolution production package has far-reaching crop protection implications. And the substitution of a few alien plant genotypes for the traditional varietal mixtures over vast geographic areas constitutes a major crop protection hazard. This condition has already contributed to numerous disease and insect outbreaks in developing countries. But the green revolution experience is so new that endemic pest population shifts in response to the changed agroecosystem are documented only rarely in the literature.

These characteristics of modern agroecosystems may suggest their abandonment in favor of the traditional, less-input-intensive production systems which are less subject to disease attack, but that alternative would not meet the world food production needs. But we must recognize the need for improved crop protection methods to reduce the level of disease loss in the modern agroecosystem and to provide greater insurance against the occurrence of catastrophic epidemics such as the Southern Corn Leaf Blight attack of 1970.

The crop protection challenge associated with modern agriculture demands that we seek to understand more fully population dynamics and interactions of crop pests within the agroecosystem, and the interactions of our independent, palliative measures to control them. Society cannot afford individual crop protection specialists making recommendations independently of and in possible conflict with actions of crop protection colleagues. Too often the farmer must integrate crop protection and production technology into a workable package at the farm level. His needs require and deserve a different approach. The complexity of crop protection tactics makes it difficult for the farmer to integrate piece-meal recommendations into a workable management tool. We must have integrated pest management teams that pursue research and extension activities on a coordinated, ecological basis in developing practical and economical management regimes for major agroecosystems. These scientists must recognize the highly complex biological and environmental interactions that

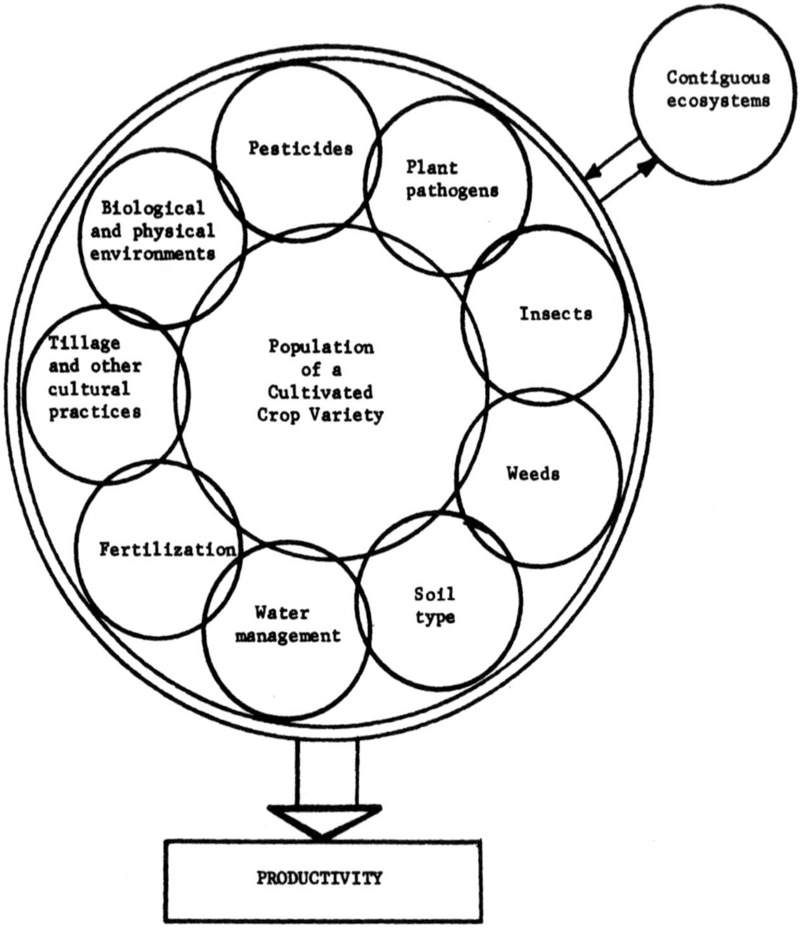

Figure 1. Conceptualization of an agroecosystem as a series of interlocking physical, biological and management functions interacting to determine the yield of a population of cultivated plants. The boundary of the agroecosystem is difficult to establish but is generally the area embracing the principal elements to be managed. Contiguous ecosystems (both natural and agricultural) will interact with an agroecosystem, but these interactions are of secondary importance. The crop pests (pathogens, insects, weeds) are important elements of the system that interact both among themselves and with the other components of the system. Crop pests are not isolated elements of the agroecosystem and are managed not as a primary objective but only as they reduce the productivity of the system by an amount greater than the cost of implementing a management strategy (economic threshold).

characterize the modern agroecosystem and seek to quantify these interactions through multidisciplinary research (Figure 1). This approach is the basis for integrated pest management and must be pursued vigorously if we are to reduce present disease losses to acceptable levels and if we are to avoid the occurrence of catastrophic disease outbreaks in both developed and developing countries.

References

1. J. E. Newman and R. C. Pickett, Science 186, 877 (1974).
2. J. L. Apple and R. F. Smith, Development Digest 10(4), 98 (1973); E. E. Saari and R. D. Wilcoxson, Ann. Rev. Phytopathol. 12, 49 (1974).
3. A. J. Ullstrup, Ann. Rev. Phytopathol. 10, 37 (1972).
4. Economic Research Service, Foreign Agric. Rpt. No. 98, U. S. Dept. of Agric., Washington, D. C. (1974) 1-98 p.
5. ________________________, Statistical Bull. 569, U. S. Dept. of Agric., Washington, D. C. (1977) 1-33 p.
6. F. Dovring, The World Food Crisis and the Challenge to Agriculture (Inst. of Nutrition, Univ. North Carolina, Chapel Hill, 1974) 1-14 p.
7. National Academy of Sciences, Population and Food - Crucial Issues (Nat. Acad. Sci., Washington, D. C., 1975) 1-50 p.
8. Agricultural Research Service, Losses in Agriculture (U. S. Dept. of Agric., Washington, D. C., 1965) 1-120 p.
9. Council on Environmental Quality, Integrated Pest Management (U. S. Gov. Printing Office, Washington, D. C., 1972) 1-41 p.
10. D. Pimentel, Bull. Entomol. Soc. Am. 22, 20 (1976).
11. H. H. Cramer, Plant Protection and World Crop Production (Leverkusen, West Germany, 1967) 1-524 p.
12. Food and Agriculture Organization, Production Yearbook (FAO, Rome, 1974). 1-325 p.
13. E. G. Leigh, Proc. Nat. Acad. Sci. USA 53, 777 (1969).
14. F. W. Preston, Brookhaven Symp. Biol. 22, 1 (1969).
15. K. R. Baker and R. J. Cook, Biological Control of Plant Pathogens (Freeman, San Francisco, 1974) 1-433 p.
16. P. R. Jennings, Science 186, 1085 (1974).
17. P. E. Waggoner, Phytopathology 52, 1100 (1962).

4

Weeds and World Food Production

William R. Furtick

The major increase in the efficiency and productivity of agriculture in advanced countries, primarily in Europe and North America, after World War II was possible only by the development of highly effective methods of weed control, together with the control of other pests. The rapid increase in productivity per acre on farms in the highly developed agricultural countries has been aided by the introduction of effective machines to work the soil, to remove previous vegetation, and to permit the timely seeding of crops over the entire farm. Crop yield has been considerably increased by the introduction of improved varieties based on the detailed work of plant breeders. Maximum productivity from these improved varieties is ensured by the use of optimum levels of fertilizer and, to an increasing extent, by improved soil environment through the drainage of excess water or conversely through irrigation. All of these ingredients for maximum production would be futile if effective control of weeds were not included in the production package.

Highly effective, selective weed control in cereal grains is a necessary forerunner to the use of fertilizer. Without selective chemical weed control, the dramatic increases in grain yields in the United States, associated with increased use of fertilizer, would not have been possible because the weeds would have responded to the added supply of nutrients. Evidence of this is often found in the less developed countries where fertilizer may be introduced without the application of modern weed killers. Where this happens, the weeds disadvantage the grain crops, drastically reducing the yield and illustrate the inadequacy of fertilizer alone as the panacea for the increased food supply so desperately needed for hungry populations. This has been dramatically demonstrated in Turkey on wheat by the

USAID/Oregon State contract team (Figure 1).

Figure 1

Summary of Weed Control Trials in Wheat Fields in Turkey--USAID/Oregon State University Team 1975 Report

In several demonstration fields a control area was left untreated to compare with the treated fields. Controlling weeds with herbicides increased the yield kg/da.

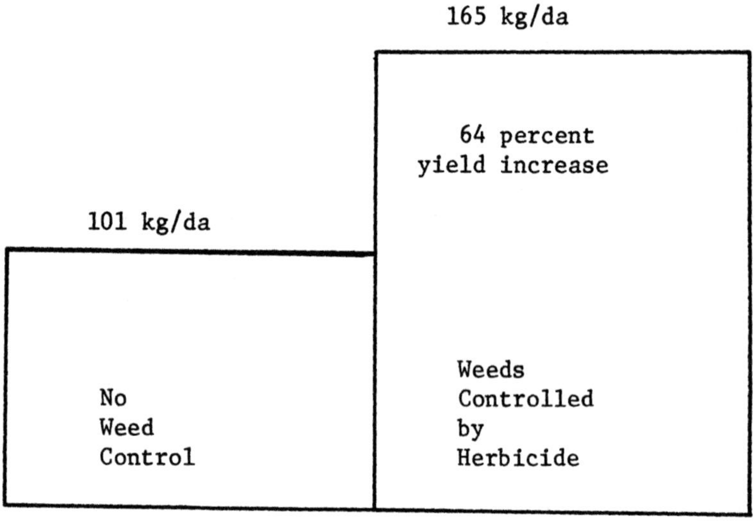

Average yield from strips with no weed control compared to yields from the same fields where weeds were controlled with herbicides.

Similar results on upland rice has been shown by the International Rice Research Institute (IRRI) in the Philippines and the International Institute for Tropical Agriculture (IITA) in Nigeria.

The loss of crop production from weeds is high in countries with the most advanced power equipment and chemical herbicides, the total losses in world food production and in other economic areas cannot even be estimated.

These losses are caused by factors which are outlined:

A. Direct Competition for Water, Light, and Nutrients

The most obvious damage weeds do to crops is the direct competition for the essential requirements for plant growth. These comprise, in addition to suitable temperature and aeration, an adequate supply of water and mineral nutrients that enable the plants to make maximum utilization of light energy for photosynthesis and to produce the ingredients for growth. When any of these essential elements becomes limiting, growth is reduced or ceases. The immense diversity of plant species combined with their adaptability to virtually any circumstances through such mechanisms as seed dormancy, perennial reproductive organs in the soil, highly efficient means of seed dispersal, and similar methods of insuring the continuity of the species, makes weed competition a certainty. Much of the world's land area is either arid or semiarid in climate, so that available moisture is one of the major limiting factors to maximum plant growth. Much of the world's cereal acreage is raised on soils that do not receive adequate moisture for maximum crop production, and therefore the water consumed by weeds under these circumstances subtracts from the yield of the grain. In fact the loss is even more serious than a simple subtraction of the weight of weeds from the weight of the grain produced since the edible portion of many crops, particularly grains, is the seed which is only the endpoint in the growth cycle of the plant. If moisture is inadequate for the plant to complete this growth cycle, then it is the human food portion that is most seriously reduced. Where moisture is limiting, grain crops wither and any grain produced is a shriveled fraction of the original potential yield.

The significant response to fertilizers, found in virtually every area of this country and in the world in general, illustrates that although many soils are abundantly endowed with the basic nutrients needed by plants, such elements are seldom all present in adequate and balanced supply to effect maximum growth. Since weed species are highly effective in competition with crops for the available nutrients, it follows that those elements in most limited supply must be shared between the weeds and the crops, and as the former are characteristically more vigorous, obviously crop growth cannot attain its maximum in the struggle for these essential nutrients.

It is well-known how plants use light energy through photosynthesis to generate the ingredients needed for growth. Thus plants growing in dense shade are generally spindly and weak. Such shading necessarily occurs when crop plants have their supply of sunlight cut-off by an overtopping canopy of weeds.

The difficulty in producing good crop yields despite the competition of weeds by attempting to supplement the growth requirements to sufficient levels for both the crop and the weeds is impossible due to the interrelationship of the several growth factors involved. Thus, for example, if the nutrient supply is made adequate for both the crop and the weeds by adding fertilizer, then the consequent abundant growth of both crops and weeds merely depletes the available moisture the more rapidly. If, in turn, this is corrected by applying irrigation to meet the water deficiency, then those weeds with tall growth potential stimulated by the adequate supply of water and nutrients will promptly overtower the crop and limit its growth by shading. Numerous studies have shown that under a given set of environmental circumstances such as amount of water, available nutrients, and the climatic conditions of a given season, an acre of land can produce just so many pounds of total vegetative dry matter in the forms of roots and top growth. Clearly then, in order to obtain maximum crop yield, all of this growth should be crop, for any weeds simultaneously growing on this acre of land subtract from the total vegetative potential and represent loss from the crop production.

B. Loss of Crop and Livestock Quality

Although the major loss from weed competition is the reduction in yield, an additional little-recognized loss is due to the more subtle effect of lowered quality of product. Thus the competition of weeds for water, for nutrients, and for light inevitably results in a lowered quality as well as yield of crop. In grain crops this often takes the form of light, shriveled grain with poor milling quality; in vegetable crops it results in small and usually malformed produce. The housewife rejects small, knobby, gnarled potatoes and spindly, twisted carrots. Similar losses in quality as decreased size, smoothness, juiciness, and uniformity occur with virtually every fruit and vegetable crop cultivated. Even the taste of such impaired produce is often spoiled, since the slow growth induced by severe competition from weeds may elicit a strong flavor in such crops as onions, radishes, lettuce, celery, and others grown largely for their contribution of taste to our food.

Weed infestation among field crops also creates economic loss through reduced quality and through the need to restore this quality in the marketable product. For example, abundant succulent weeds growing in a grain crop often add moisture to the grain at harvest, necessitating considerable additional expense for drying to prevent molding and spoilage in storage. Also, contaminating weed seed, not readily cleaned from the grain, can render such grain unfit for milling. Thus, many seeds, of similar weight and density to that of grain, such as vetch, wild garlic bulblets, bindweed seed and many others will give flour an off-flavor or, in some cases, may even be toxic for human consumption.

The contamination of seed crops with weed seeds is understandably a major concern to all crop agriculture. For when a grower sows seed for a new crop, he does not want to plant with it seeds of a serious weed species, many of which may not have previously been present on his farm. This has led to legislation relative to the control of the weed seed content in marketed seeds in the United States, which necessarily resulted in increased cost for the production of weed-free seed or nursery stock.

Weeds also create problems in the livestock industry because they often contaminate hays, thus reducing their quality as livestock feed or causing feed efficiency losses because the animals refused to eat such contaminated forage. Where the weed contaminants of either dried hays or open pastures include poisonous plants, then direct losses from death or sickness of foraging livestock may result, and there are few pastures that do not contain at least a few such poisonous weedy species. Although farm animals that graze on poisonous plants may not ingest sufficient amounts to cause death, resulting sickness will reduce their growth and they will fail to fatten or to produce the milk desired, thus effecting losses both in quantity and quality of livestock product. Such obscure losses are probably much larger than is generally recognized. Again, many weed species, though not poisonous to livestock, do contain substances that impart a bitter or otherwise strong off-flavor to milk; a good example of this is cows grazing pastures infested with wild garlic will yield milk that tastes of garlic.

Many weeds produce burrs or thorns that become entangled in the hair of sheep and goats, reducing both the quantity and quality of the wool or hides.

Large areas of the world's rangelands are infested with unpalatable weed species that substantially reduce the grazing potential, thereby causing large losses both in meat production and in feed needed to fatten the marketable livestock adequately.

C. Weeds as an Intermediate Host for Insects and Diseases

Many of the insects and disease organisms that attack crop plants spend part of their life cycles on various weed species, where they thrive and multiply, and build to a population density that can damage crops. A variety of examples can be cited to illustrate this relationship. In some cases weeds also are a benefit by providing a reservoir of predators and parasites that are important in biological control programs for crop pests.

D. Weeds Reduce the Value of Land or Prevent Land Use

Because of the immense difficulty in controlling many species of particularly aggressive perennial weeds, infestations with certain of these species can seriously reduce land values. These are weed species that have aggressive growth habits and vegetative reproductive system. Some of the more common of these species native to the United States include Johnson grass (*Sorghum halepense*), quackgrass (*Agropyron repens*), Canada thistle (*Circium arvense*) bindweed (*Convolvus arvensis*), Russion knapweed (*Centaurea repens*) whitetop, (*Cardaria draba*), leafy spurge (*Euphorbia esula*), and nutsedge (*Cyperus* spp). Land infested with such perennial weeds has a reduced sale value simply because a prospective buyer is hesitant to pay full price for land, the agricultural potential of which is seriously limited. In extreme cases, where land is totally infested with one or more of these serious perennial weeds, its agricultural utility may be so completely reduced that it is just abandoned. In some tropical areas of the world the

Cyperus sp. and *Imperata* sp. render large areas of very fertile land non-productive.

E. Weeds Increase Production Costs

One of the major costs in crop production is the energy required for soil tillage and cultivation, at least half of which is devoted solely to weed control. This aggregates annually in the U.S. a total of more than 60 million horsepower. It is estimated that much of the energy is utilized merely to move 250 billion tons of soil each year, an amount of soil that would comprise a ridge 100 ft. high and one mile wide from New York to San Francisco. This is a really staggering exertion when one considers that more than half of it would be eliminated except for the need of weed control. In addition to the expense for machinery used for tillage and cultivation, weeds also become a major economic factor in the operation of harvesting equipment; for weedy fields clog potato diggers, slow and stall corn pickers, require special design and careful setting of combines to separate grain and seed crops from their straw and weeds and, in many other ways, interfere with efficient harvesting. Weeds substantially increase the costs of processing because special equipment is required for cleaning weed seeds from seed crops, such as separating weed seeds from grain and removing weed contaminants from many crops, including such processed food crops as vegetables. A good example of this unnecessary expense is the hand removal of Canada thistle flower buds and similar weed parts from peas under process for canning or freezing.

In addition to all these costs, there is the direct expense of weed control apart from that of the regular tillage and cultivation. These costs include not only the considerable investment in hand labor to remove weeds from intensive crops, such as vegetables, cotton and bush fruits, but also the price of millions of pounds of herbicides applied each year plus the cost of their application.

The Role of Weeds in Crop Production Systems

In the history of the development of agriculture, an awareness of the magnitude of impact weeds have on food production has been slow in developing. Part of this has probably resulted from the fact that weeds have always been an integral part of the crop ecosystem and since they do not bite or blight the crop, their damage has been too subtle to arouse concern. It has long been known that some crops have reasonable production capacity in spite of weeds while others do not. From the earliest stages of agriculture, production systems have been developed to allow crops to compete adequately with weeds to insure reasonable yield.

One such method was to plow under the weed seeds and debris from previous crops, deeply enough to prevent easy germination and growth of weed seedlings; this would give the crop a maximum chance to thrive against seedlings from the remaining viable seeds buried in the previous plowing. Cultivators were also developed to eliminate weeds from between the rows of such crops as corn, beans, cotton, and potatoes. One of the earliest weed control measures practiced was crop rotation, which precluded the excessive buildup of those weeds well adapted to a particular cropping practice. This was gone simply by rotation to a different crop or practice that was detrimental to the weed species that had proliferated under the previous practice; by continually rotating crops in this manner, no one weed species could attain a high level of infestation. In the more primitive agriculture, hand pulling weeds is still the primary means of insuring crop growth.

Mowing weeds to prevent their seeding and to establish perennial crops is another long-practiced method of weed reduction, particularly in the establishment of legumes and forage crosses. These are generally small-seeded species that have a relatively weak seedling and cannot compete with tall, rapidly growing weed species; frequent moving permits these small-growing seedlings to become established and, once they are vigorous perennials, to dominate the weeds in future years.

Many of our present cropping practices are themselves determined by the type of weed control methods necessary for efficient production. Actually, the current method of a particular crop culture may not be the best to employ were weeds not a major production problem. A good example of this dependency is the spacing of rows used for various crops. Those crops capable of rapid growth and vigorous competition with weeds are frequently sown broadcast or in closely drilled rows; typical of such crops are the cereal grains. Conversely, many slow-growing vegetable crops traditionally are spaced in wide rows far enough apart to permit easy cultivation or hoeing to eliminate weeds. Although it has long been recognized that cereal grains are capable of high production efficiency when sown in closely spaced rows or broadcast, few realize that so also are carrots, which generally are raised only in spaced rows. Now that highly effective herbicides are available for removing weeds in carrots selectively, carrot culture is being reappraised to explore the possibilities of increasing production efficiency with broadcast or closely drilled rows.

Unfortunately, it has only been during the last two decades that major attention was focused on the interaction of various weed populations at various stages of development in the different types of crops to carefully measure their impact on production. This type of research was fostered by the increased crop growth and health observed when newly developed selective pre-emergence herbicides were introduced which permitted crops to germinate and grow in nearly a weed free environment. These discoveries coincided with the rapid expansion in use of high fertility levels on new crop varieties bred for their capacity to respond to fertilizer. Weeds caused greater problems than when lower yields were the norm without fertilizer.

The results of careful crop-weed ecosystem competition research in many research programs in various parts of the world during the 1960's and still continuing has demonstrated that weed competition, especially in many of the major row crops, which have become crops grown in rows because of sentivity to weed competition, occurs much earlier in the growth cycle than anticipated. Work in crops such as corn has shown that competition prior to the normal cultivation or hand weeding practiced in both modern and traditional agriculture already has substantially reduced yield potential. This means that where pre-emergence selective herbicides are not used, the early weed growth plus those left in the row by cultivation have caused 10-30 percent loss in yield potential without readily observed adverse effect on crops. The increased appearance of health for crops treated pre-emergence with selective herbicides during early research with these compounds lead to initial speculation that these new chemicals had some type of growth regulating or fertilizing action in addition to killing weeds. These responses are now more clearly understood as part of previously unobserved competition effect which were now eliminated.

Wide Scale Use of Herbicides in Modern Agriculture

During the past few years a wide array of selective herbicides have been developed and marketed which can give a reasonably weed free environment for most major crops. This has been required in order to use present monoculture cropping. In the highly developed agricultural countries, the use of herbicides has become as basic to crop production as the use of synthetic fertilizer nutrients. This has resulted in herbicides, next to fertilizer becoming the large tonnage volume sector of the agricultural chemicals industry. Even though herbicides are classed as pesticides and regulated as such from a legal standpoint, they now more closely resemble fertilizer from a use viewpoint. The nearly universal use of herbicides wherever fertilizer is applied to crops is due to the substantial yield loss associated with failure to control weeds. Unlike plant diseases, insects and similar pests, weeds are a dependable unwanted part of the cropping system every year.

The types of crop production practiced tend to determine the weed species that dominate a given area. Those weeds best adapted to the pre-dominant culture build up and dominate, while those the least adapted disappear. An example of such changes is illustrated by those annual weed species that infest wheat, depending on whether winter or spring wheat is produced. In winter wheat production areas, winter annual weeds dominate, while with spring wheat, the summer annual weeds predominate. Each crop and each cultivation practice has its own distinctive weed problems. With the great diversity of weed species, it generally happens that a number of species fit into any given type of environment so that weed problems frequently are not confined to a single species but involve numerous species that become adapted to the particular situation. This means that, for the most efficient weed control, field practices must be designed to combat a wide array of different weed species. Frequently, control measures actually tend to induce changes in weed species rather than to eliminate them entirely. Those weeds sensitive to the specific control measures employed disappear, while those not so adversely affected merely move in to take their place.

The number of plant species that, under certain conditions, may act as weeds is extremely large. Many species are commonly found to be serious weed problems throughout the world; some are localized to a few geographic regions;

while others may be actually both crop plants and weeds, depending entirely upon the circumstance of where they are growing.

Looking at it from a global impact on food production, we can then summarize this discussion by indicating that weeds are a complex part of the crop ecosystem that require constant manipulation to keep their adverse impact on crop yield to a minimum. In modern agriculture, use of selective herbicides has recently become a nearly universal regulating factor of this competition. The late awareness of the role of weed control in minimizing food production losses has created serious underfinancing of weed research outside the agri-chemicals industry. There is an increasing diversity of research on various non-chemical methods, using everything from biological means to high frequency sound waves. There is urgent need for greater investment in these types of research. Most of the agriculture in the world is still not in the modern sector and much of it in the tropics where year around growing conditions and climate favor high density and rapid weed growth. Very few of the less developed countries have any effort specifically devoted to training, research or extension dealing with weed control in contrast to plant nutrition, pathology and entomology. This must be corrected if modernization is to occur.

Even in the highly developed world, weed control research is indicating that many of the long traditional agricultural practices that have been handed down from early agriculture that were based on weed control needs must be re-examined. These are such things as row spacing, soil tillage, cultivation, etc., the changing need for food and energy conservation mandate speeded up effort to re-examine traditional cropping practices. With the current resurgence of interest in agricultural research, weed control must be kept in mind as a basic component. The fact that herbicides are classed as pesticides and not fertilizer in spite of use characteristics means care must be exercised not to strangle their research and development under a mass of government regulation. Let us hope that reason and awareness will help insure less loss in food production from weeds in the future.

References

1. Plant Protection and World Crop Production. Bayer, A.G., Leverkusen, Germany, 1967.

2. Weed Control by Chemical Methods. Matthews, L. J. A. R. Sherer, Government Printers, Wellington, New Zealand, 1975.

3. Weed Control, 3rd Edition. Alden S. Crafts and Wilfred W. Robbins, McGraw-Hill, 1962.

4. Weed Science Principles. Wood Powell Anderson, West Publishing, 1977.

5. Weed Science Principles and Practices. G. C. Klingman, F. M. Ashton and L. J. Noordhoff, John Wiley & Sons, 1975.

5

Animal Pests and World Food Production

Roger O. Drummond, Ralph A. Bram, and Nels Konnerup

Man's domesticated livestock (the source of almost all animal proteins consumed by humans) may be looked upon as direct competitors for the grains needed in the diet of humans or as providers of luxury protein items in the diets of economically privileged people; conversely, animals may be regarded as rich sources of essential amino acids, fats, vitamins, and minerals for human needs or as the only organisms that can convert cellulose in plants, mainly grasses found on nonarable land, into food for human consumption. We propose to briefly examine livestock production as it relates to human dietary needs for animal proteins, as a source of animal proteins, as a reponse to world food needs, and as a means of using the earth's resources. Detailed discussion will focus on the effect of pests, particularly arthropod pests, on the efficiency of livestock production and on the effect of livestock pest control on the environment.

Animal Protein in the Human Diet

Although plants can manufacture all the 20 amino acids found in their proteins from simple carbon and nitrogen compounds, animals must obtain about half of the amino acids they need from external sources. Humans must have at least 45 and possibly as many as 50 dietary compounds and elements for a healthy life (1). (Nutritional requirements of humans are not fully defined and are still the subject of investigation (2).) However, the "normal" human requirement is not for dietary protein per se but for adequate amounts and appropriate proportions of 9 or 10 essential amino acids. In general, plant proteins contain inadequate amounts of one or more of the amino acids essential to human nutrition; conversely, a protein source such as meat, eggs, or milk contains enough of all the essential amino acids required for human maintenance and growth.

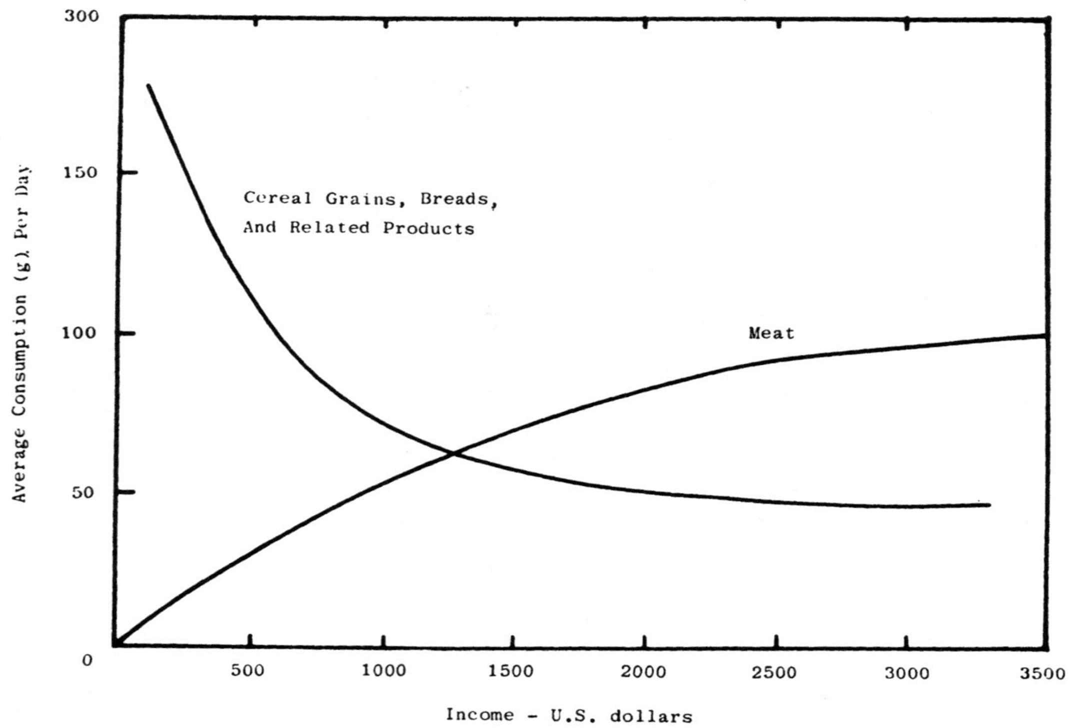

Fig. 1. Consumption of plant and animal products in relation to income (3).

Unfortunately, many persons in the world do not have adequate amounts of animal proteins in their diets; an example is the direct correlation between income and the ratio of meat and cereal products in the human diet (Fig. 1). (In many diets, meat is replaced by animal products such as milk and eggs.) Obviously, the diets of some segments of the human population are oversupplied with animal proteins, whereas the diets of other segments do not include animal proteins because of income, social and religious constraints, or other reasons (4). Nevertheless there will be greater and greater need for animal proteins in the human diet as the world population grows and the level of affluence increases in many countries.

The Source of Animal Proteins

Animals, the only source of animal proteins, have dietary needs such as carbohydrates, fats, vitamins, and amino acids. They must take all these substances directly or indirectly from green plants. However, animals vary considerably in the processes by which they digest the products of plant growth (5). One class, carnivores and omnivores (dog, man, pig), has a simple stomach and cannot adequately digest cellulose; a second class is composed of nonruminant herbivores (horse, rabbit) that digest cellulose inefficiently; and a third class, ruminant herbivores (cattle, sheep, etc.), digests cellulose efficiently. In the ruminant, enzymes of the bacteria in the rumen break cellulose into smaller molecules; then the transformation into amino acids and vitamins takes place before the food passes into the small intestine, which can absorb the nutrients.

Green plants use solar energy to convert carbon dioxide and water into carbohydrates (mostly cellulose), which, in turn, are converted to animal proteins by ruminants. When one considers the first process in terms of energy conservation, plants transform 15-20 percent of the available solar energy into chemical energy, but only about 20 percent of this chemical energy can be used directly by man through consumption of plants. Domestic livestock convert plant products (energy and proteins) to food for humans at efficiencies that range from 2 to 26 percent (Table 1).

Table 1. Efficiency of conversion of crude protein and energy of plants to edible products by animals.

Animal (product)	Conversion efficiency (percent) Energy	Protein
Poultry		
(eggs)	18	26
(meat)	11	22
Cattle		
(milk)	17	25
(beef)	3	4
Swine		
(pork)	14	14
Sheep		
(meat)	2	4

The fact that ruminants need not compete directly with man for the energy and protein available in plants is enough to justify their place in the production of animal proteins for human consumption. But, in addition, ruminants and other animals can convert other nutrients such as plant wastes, sewage sludge, manures, and by-products of plant and animal processings into animal proteins for human consumption. Also, in many developing countries, animals provide much of the mechanical energy for cultivation.

World Food Needs and Natural Resources

There is a general consensus that the population of the world, estimated now at 4.0 billion, is growing at a rate of about 2 percent per year (6). This rate is presently much higher in developing countries (Fig. 2) though it is decreasing slightly. The increase in human population will create an increased demand for food, which includes additional needs for both plant and animal protein. However, animal proteins are more costly to buy than plant proteins, and their production is much more expensive in terms of land, energy, fertilizer, and water (Table 2).

Table 2. Resources necessary to produce one pound of protein from plants or from beef (7).

Resource (unit)	Amount needed for indicated source Plant	Beef
Land (acres)	0.05 - 0.13	0.16 - 0.41
Energy (million BTU)	.1 - .8	1.8 - 12.7
Fertilizer (lb)	2.0 - 13.0	8.0 -158.0
Water (acre/feet)	.03 - .07	.29 - 1.34

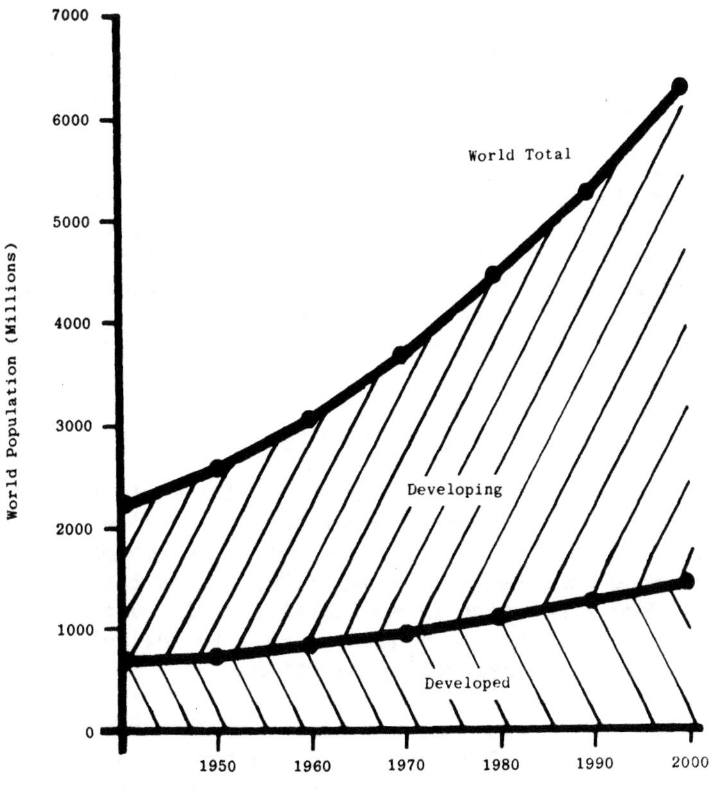

Fig. 2. World population growth; past and projected (3).

The question is then whether the world can afford the plant production and energy necessary to increase the production of animal proteins. Also, what resources are available and what additional ones are needed if this production is to meet minimum human dietary needs?

The natural resources of our earth are finite (8). Of the total land surface of 13 billion hectares, over 4.5 billion hectares cannot be cultivated because of climate, and an additional 5.3 billion hectares (mountains, deserts, and poor soils) are not suitable for cultivation. This leaves only 3.2 billion hectares (24 percent of the land surface of the earth) that are arable, but in any one year, only one-half to one-third this area is actually harvested. In order to make all arable land productive, it would be necessary to make tremendous expenditures of water, soil, fertilizer, technology, and funds. Meanwhile, 3.6 billion hectares of the 5.3 billion that are not suitable for cultivation are suitable for grazing livestock because this land supports grasses and other plants not suitable for human food. Ruminants can convert these plants into highly suitable proteins. Ruminants and other animals can also convert many of the wastes from plants grown on arable land into human food.

The Effect of Pests on Animal Production

The maximum use of the resources of the world to provide food for the increasing world population is a prime goal of both developed and developing countries. The conferees at the World Food Conference in Rome, 1974, agreed that increased food production is essential in all countries and that all resources must be made available in maximum quantities in order to realize full production.

The law of the minimum, formulated by Justus von Liebig in 1840, states that plant growth is limited by the availability of whatever nutrient is scarcest. The application of this law in terms of basic ecological principles and concepts has been expanded by Odum (9). But if Liebig's law is limited to essential factors such as nutrients or chemical materials necessary for physiological growth, then there is a need for other concepts that take other factors into consideration. Shelford's law of tolerance states that absence or failure of an organism in a given ecological system is controlled by the qualitative or quantitative deficiency or excess of any of several factors (physical, biological, etc.) that approach the limits of tolerance of the organism. These concepts were combined into Odum's law of limiting factors, which states that because the presence *or success of an organism or group of organisms depends upon a complex of conditions, any condition that approaches or exceeds the limits of tolerance of the organism is a limiting factor or condition. This law of limiting factors has direct application to pests and animal production because production of animals can be maximized only if the following three factors are present in amounts that are not limiting: (1) adequate food, including water, grass, minerals, and other similar items of the environment; (2) a suitable breed of animals that has known tolerances and requirements; and (3) sufficient management (Fig. 3). Management includes a variety of acceptable animal husbandry practices such as pasture rotation, supplemental feeding during stress periods, mixing of sexes for breeding at specific times, and especially attention to animal health. Attention to animal health is used in a general sense to include well-being of animals despite the many diseases and pests that continually threaten to decrease or eliminate productivity.*

Animal health can be separated into two general divisions, diseases and pests. The major diseases that limit livestock production include those of international significance that are listed in Table 3.

Table 3. List of some important diseases of animals (10).

Epizootics	*Enzootics*
Rinderpest	*Wide distribution*
Contagious bovine pleuropneumonia	*Brucellosis*
Hemorrhagic septicemia	*Tuberculosis*
Foot and mouth disease	*Anaplasmosis*
African horse sickness	*Mastitis*
African swine fever	*Vibriosis*
Hog cholera	*Lumpy skin disease*
Newcastle disease	*Bluetongue*
Fowl plague	*Limited distribution*
Trypanosomiasis	*Ondiri disease*
East Coast fever	*Nairobi sheep disease*
Piroplasmosis	*Rift Valley fever*

Diseases are obviously of major importance to the production of livestock. However, figures for losses due to the major epizootics and enzootics (Table 3) often include losses due to the reactions of animals to infestations of arthropod and helminth pests. For example, losses in animal production caused by animal death, destruction of animal products, and decreased production have been calculated to average 17.5 percent in Europe, Oceania, and North America and 35 percent in the rest of the world (11). Also, losses caused by animal

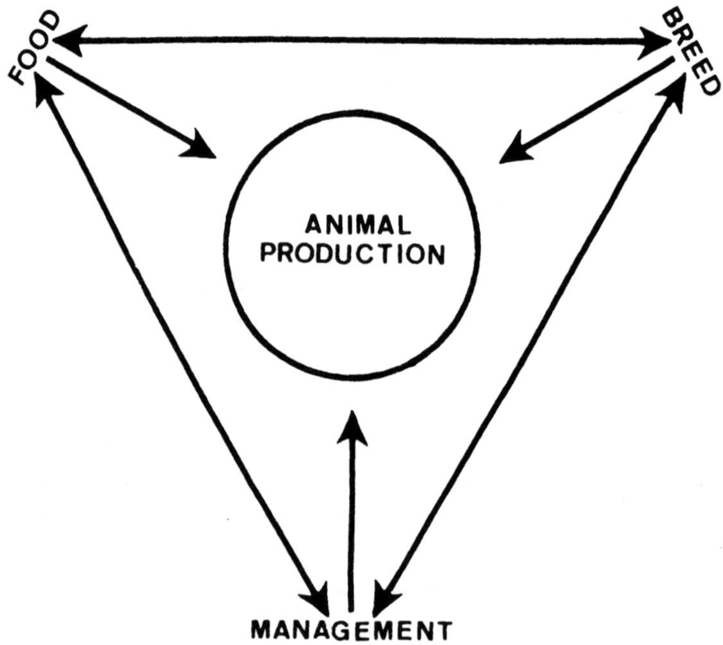

Fig. 3. Interaction of factors that affect animal production.

diseases have been estimated (12) to range from 1 percent of the total production to as high as 37 percent; but pests such as parasitic helminths and arthropods are responsible for 30 to 91 percent of the losses caused by these diseases.

Epizootics listed in Table 3 are major factors in limiting livestock production in some countries (13) and are therefore objects of international quarantines and studies designed to prevent their spread and to eventually lead to their cure (14). Although diseases are thus of great significance to animal health, they will not be further reviewed here except as they are transmitted by an arthropod pest.

Because pests are important in animal management, the presence (or absence) of a pest is a factor that is critical to the maximum production of livestock. Throughout the world, infestations of pests are present in livestock, and these pests can impact seriously on both production and the economy in given areas (15, 16, 17). For example, the tsetse fly is a prime limiting factor in that part of the world where it is found because it essentially prevents livestock production. In other situations, pests reduce productiveness despite adequate food and a potentially high-performing breed.

Basically two types of pests affect livestock: (1) helminths--worms that live in the lungs, gastrointestinal tract, liver, and other body organs of animals and (2) arthropods--insects, ticks, and mites that are parasitic on animals.

Helminths will be dealt with only briefly, but their presence can cause a variety of pathological effects in animals (18), and they are of significance in the raising of livestock (19). The effect of helminth infestations on the production of wool in sheep (20, 21) and on weight gains and deaths (22) has been well documented and justifies the strategic use of anthelmintics to avoid less than optimum productivity. However, the age of sheep at time of treatment, the movement of treated sheep to worm-free pastures, and factors such as weather can influence the size of worm burdens that directly affect the animals (23). The effect of gastrointestinal helminths on the production of beef and dairy cattle is less well established. In some studies (24, 25), treatment (anthelmintics administered frequently in the feed, by injection, or by drench) has had no effect; in others, there have been increased weight gains as high as 73-74 pounds per calf (26) or 0.15-.18 pounds per average daily gain in field and feedlot trials (27). Also, gains of up to 150 pounds per calf resulted from one strategic drenching and subsequent movement to worm-free pastures (28). In general, such treatments do not adversely affect the environment. However, it is extremely important that anthelmintics not be used indiscriminately so as to prevent, delay, or reduce the enhancement or acquisition of natural host immunity to these parasites (29). Host immunity offers a more permanent and thus more satisfactory solution to the problem of losses in livestock due to helminths.

The second type of pests affecting livestock, arthropod pests, is the subject of the remainder of this report. We will present a limited overview of the effect that arthropod pests have on animal production and, in turn, the effect that control of these pests (by chemical pesticides and other means) has on the environment.

Although animals are attacked by a variety of arthropods (Table 4), a relatively few species can be classified as being the factor that limits animal production.

Table 4. Arthropod parasites on domestic animals in the United States, its possessions, and Canada (30).

	Numbers of species in indicated group				
Host	Mites	Ticks	Lice	Flies*	Fleas
Cattle	7	22	6	10	0
Sheep and goats	7	13	9	10	1
Horses, mules, and asses	5	16	2	8	1
Swine	2	10	1	5	3
Birds+	23	6	38	1	4

* Does not include such flies as house fly, stable fly, etc., that are usually not collected on the host.

\+ Includes chickens, turkeys, pigeons, pheasants, ducks, and geese.

Also, though there is a great deal of literature on losses due to arthropod pests (losses are generally considered high), accurate, well-substantiated data on economic thresholds are lacking. Thus, there has been limited analytical quantification of these losses and the resultant need for or benefits of the application of insecticides to control these pests (31).

As Table 4 shows, sheep and goats are infested with 40 species of arthropods. Infestations of sheep keds, Melophagus ovinus (L.), can cause a decrease in weight gain, wool growth, and value of leather of sheep (32, 33). In addition, in Australia, losses in excess of 28 million Australian dollars (34) are reported to result from infestation of sheep, especially Merinos, by larvae of Phaenicia (= Lucilia) cuprina (Wiedemann). These larvae have been controlled by application of insecticides to sheep, but unfortunately P. cuprina has developed a high level of resistance to several organochlorine, organphosphorus, and carbamate larvicides (35). This resistance will certainly limit the usefulness of insecticides as a continuing treatment for fly strike (= attack by P. cuprina) and will create a need for other methods of control such as a breed of sheep that is not susceptible, increased use of mules operation, midseason crutching, and better animal husbandry (36). The use of genetic manipulations to aid in the control of this species is an active research program (37).

In the United States, 45 species of arthropods are reported (Table 4) as pests of cattle. Lice--both biting lice, Mallophaga, and sucking lice, Anoplura--can cause losses in livestock; for example, infestations of the shortnosed cattle louse, Haematopinus eurysternus (Nitzsch), can cause anemia and increased winter weight loss (up to 81 pounds per heifer or 55 pounds per bull) in moderately to heavily infested animals (38, 39, 40). There is a relationship between nutritional status of animals and infestations of lice in that poorly fed animals have higher louse populations (41, 42), but this relationship is not well defined. Also, there are limited data concerning the relationship between louse populations and effects on animals, or on thresholds of infestation that justify application of insecticides to cattle to control lice (31).

Flies, a major group of insects that affects livestock, can be divided into three general categories: (1) those that are a nuisance as adults but do not suck blood, (2) those that do not affect livestock as adults but have a larval stage that invades the tissues of livestock--a condition known as myiasis, and (3) those that suck blood from livestock.

High infestations of nuisance, nonbiting flies such as the house fly, Musca domestica L., around barns, feedlots, and other animal holding facilities may annoy animals to the extent that they do not gain weight or convert feed at the maximum potential level. Also, house flies are known to carry organisms that cause diseases of livestock. House flies can be controlled by treating areas where they rest. Another nonbiting fly of special interest to the United States rancher is the face fly, Musca autumnalis De Geer. This fly, which is attracted to the moist areas of the mouth and eyes of cattle and is found in almost all of the contiguous states, interferes with normal grazing and also has been associated with the increased incidence of pinkeye (infectious bovine kerato conjunctivitis) in cattle. Pinkeye was shown to cause a significant decrease in weight gain, especially in young cattle at weaning, which were lighter by 33-40 pounds (about 8 percent of body weight) (43, 44). Available insecticides, which are often applied to the cattle, are relatively ineffective against the face fly. Dung beetles that destroy and bury the manure pats that are the site of larval breeding of face flies and other species of flies before the larvae can complete development show promise as biological control agents against such dung-inhabiting Diptera (45).

Among flies that produce myiasis in livestock, the warble flies Hypoderma lineatum (de Villers) and H. bovis (L.) are known to do considerable damage to cattle. The effect of hypodermatosis on blood systems and body weight of cattle has been well documented (46, 47); and reviews of the literature (48, 31) indicate that cattle treated with systemic insecticides to eliminate cattle grubs have generally greater average daily gain than untreated cattle. Other benefits that accrue from controlling cattle grubs are (1) the increase in value of hides (those hides with no warble holes are grade no. 1 and bring $0.50-1.00 more per hide than those with five or more holes); (2) the decrease in the amount of trim of meat from cattle at slaughter (losses of over 1.1 kg. of flesh per animal due to infestation with warbles have been reported (49)); and (3) the decrease in the loss of milk and meat production that results from the gadding of cattle in the spring-summer by heel flies, the adult stage of the cattle grub.

The development and widespread use of highly effective animal systemic insecticides that can be administered orally or as sprays, dips, or pourons to grub-infested cattle have made it possible to control cattle grub larvae before they finish their migration in the body of the animal and thus can prevent losses in hide value and excessive trim. Unfortunately, the increased value of the hides and decreased trim losses are not used as a basis for higher prices for cattle that have been treated by the producer. It was to obtain such economic gains in grazing, milk flow, slaughter value, and hides that Ireland undertook a warble fly eradication campaign in which systemic insecticides were used to treat all the cattle in the country. Through 1974, more than 32 million cattle have been treated in this campaign since its start in 1965; and it is considered an outstanding success though eradication had not been achieved at latest report. However, in April and May 1975, incidence of warbled cattle was only 0.18 percent (50).

Another myiasis-producing fly that has been the subject of an eradication campaign is the primary screwworm, Cochliomyia hominivorax (Coquerel). This fly lays eggs around cuts and wounds on warmblooded animals, and larvae infest the wounds. When these larvae are not treated with an effective larvicide, the infestations often lead to the disability or death of the animals. Stockmen in the southeastern and southwestern States had to continually treat their animals during warm weather to prevent such losses. In addition, though screwworms do not diapause and are killed by cold weather, they do migrate northward during the summer and could be found in states as far north as Iowa, Illinois, South Dakota, and New Jersey. (Figure 4 illustrates the actual summer migration and winter survival in 1951-52.) In the screwworm eradication program in the southeastern United States, which was the culmination of research by the USDA that started in 1913 (51), treatment of wounds to control natural populations of these flies was combined with the release of millions of flies sterilized by exposure to radiation. The program was completely successful: the species was eradicated after a campaign of only 2-1/2 years at a total cost of $11 million. Savings are estimated at $20 million per year. In the Southwest, the eradication effort that started in 1962, though many times larger (billions of flies have been reared, sterilized, and released in the U. S. and Mexico), has not been completely successful. Screwworms were eliminated from overwintering areas in the United States, but flies migrating into the region from Mexico have been a source of constant reinfestation in Texas, New Mexico, and other border states. Nevertheless, the program has been extremely beneficial in savings of labor and insecticide and in decreased disability and death loss in livestock and wildlife. Estimates of cost per benefit ratio range from 1:39 to 1:113. The projected opening of a new fly production plant at Tuxtla Gutierrez, near the Isthmus of Tehuantepec in Mexico, will provide more sterile flies (estimated 300 million per week) for an enlarged campaign to eradicate flies in Mexico north of the Isthmus and thus to keep the U. S. free of screwworms (52).

The group of flies that suck blood from livestock, so-called biting flies, includes many species of flies and mosquitoes. The effect of these species on livestock is quite variable. There are well-documented reports (31) of extensive losses in domesticated animals due to massive feeding attacks by black flies (Simulidae) that caused death and lowered production. Black flies also cause severe dermatitis and transmit onchocerciasis and viruses of livestock. Biting gnats (genus Culicoides) transmit the virus that causes bluetongue disease of cattle and sheep and also the filarial nematodes that are pathogenic to cattle and other animals in

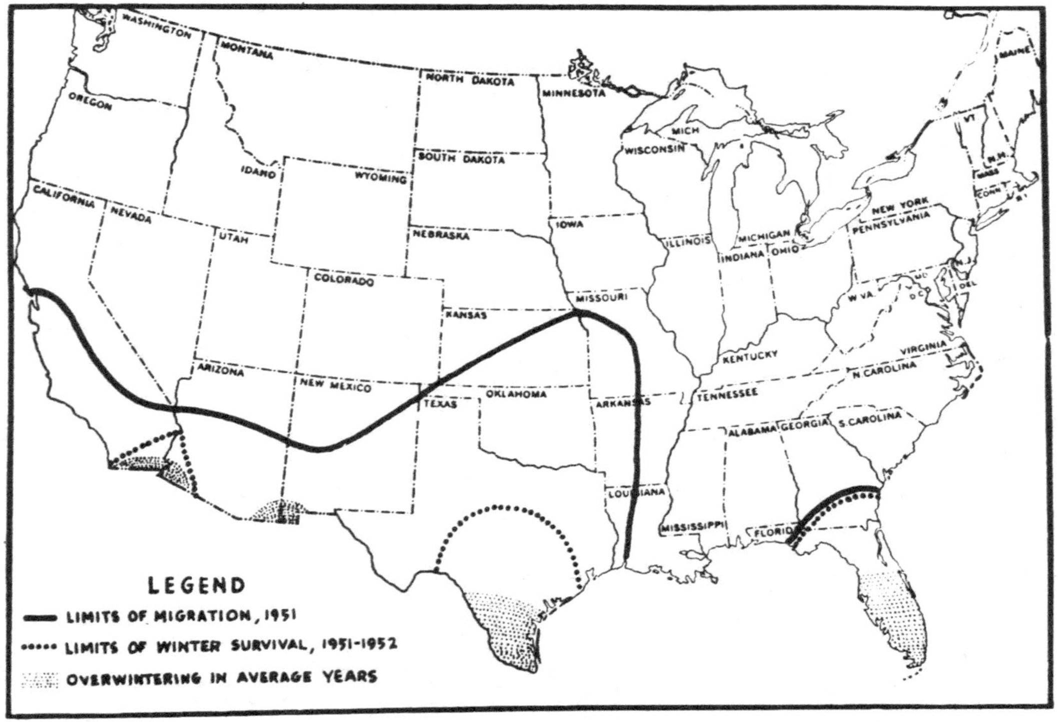

Fig. 4. Example of winter survival and summer migration of screwworms in the United States.

many areas of the world. Generally, for control of these small bloodsucking flies and mosquitoes, the larval habitats--streams, ponds, and other bodies of water--must be treated with insecticides. The adult stage can be killed only by treating large spaces. However, developing resistance to pesticides and environmental considerations mean that repeated use of insecticides is not the final solution. Other more permanent, less polluting, and safer measures must be combined into an integrated pest-management system for control of these arthropods (53).

Another biting fly, the horn fly, Haematobia irritans (L.), is found on cattle by the thousands everywhere in the United States. Horn flies can cause a reduction in weight gain and milk production of infested animals. Control has resulted in gains of 2.1-3.6 kilograms per treated animal over an 82- to 93-day grazing period (54) and has made it possible for treated cows to wean significantly heavier calves (55). Horn fly control consists of treating animals dermally with insecticidal sprays, dips, backrubbers, dusts, and pourons for the control of adult flies or treating cattle orally with larvicides so as to kill developing larvae in fresh bovine manure. Repeated applications were necessary to control horn flies, but recently, the release of sterile males and the administration of an insect growth regulator in the drinking water of cattle suppressed a population of horn flies on a ranch in Hawaii (56).

The attack of small numbers of another large, ubiquitous bloodsucking fly, the stable fly, Stomoxys calcitrans (L.), has been shown to reduce milk production by 40-60 percent, but the milk flow of a few dairy cattle with access to unlimited quantities of high-energy feed was not adversely affected by larger numbers of stable flies, though the efficiency of feed conversion was not measured (57). The control of stable flies centers around sanitation since larvae breed in decaying manure or in contaminated hay, feed, or straw. If areas around feedlots and stables are kept free of breeding medium, the population of stable flies will be greatly reduced. Treatment of livestock with repellents or insecticides generally is unsatisfactory as a control for stable flies because the treatments are either ineffective or, at best, have only short residual activity. The sterile-male technique is being evaluated by the USDA as part of a pest management system to control stable flies.

Another stable fly, Stomoxys nigra Macquart, has been described (58) as the major limiting factor in the development of a dairy industry on the island of Mauritius. There, almost all milk cattle are housed continually in darkened

Fig. 5. Africa south of the Sahara. Stippled areas are infested with tsetse fly (62).

huts to protect them from the bloodsucking and irritation of S. nigra. However, milk production that averaged 3.14 liters milk per day per cow when cows were housed in huts increased to 11.9 liters per day per cow when the same animals were held in light, airy conditions.

In a series of well-executed tests in Louisiana (59, 60, 61), mosquitoes, mostly Psorophora confinnis (Lynch-Arribalzaga), Anopheles quadrimaculatus Say, and A. crucians Wiedemann, significantly (and economically) reduced the gain (0.05 kilogram per animal per day) of feedlot steers. Purebred Brahman steers were more tolerant of mosquito attack than purebred Hereford steers, and animals on high-energy ration tolerated more mosquitoes than animals on low-energy rations. Mosquito control can include larviciding as well as adulticiding, but usually mosquitoes are controlled only when they are a problem to humans--not livestock. Additional data on cost-benefit ratios need to be obtained before mosquito control for cattle only can be attempted.

Horse flies, the Tabanidae, a family of flies found throughout the world, cause losses in animal production as a result of their blood-feeding activities. These flies are exceptionally difficult to control because of their intermittent attack and wide-ranging flight activities (31).

Of all arthropods, the tsetse fly is the foremost example of a pest that prevents the production of animal protein. The tsetse fly is undoubtedly the most formidable deterrent to the production of cattle in an area of 7 million square kilometers (4-1/2 million square miles) of savannah in the central area of the African continent (Fig. 5). All of the 30 species or subspecies of Glossina feed on vertebrates and transmit trypanosomiasis, a disease caused by the protozoans Trypanosoma congolense Broden 1904, T. brucei Plimmer and Bradford 1899, and T. vivax Ziemann 1904, that is fatal to livestock not tolerant to the disease. As a result cattle production is limited to animals that are trypanotolerant or to susceptible cattle that are raised under special ecological conditions or are given regular trypanocidal treatment. The fact that cattle cannot be raised where tsetse exists is amply illustrated by the distribution of cattle and presence of tsetse in Tanzania (Fig. 6). The socioeconomic consequences of the failure to raise a possible 120 million cattle because of tsetse include lower human nutrition, reduced general agricultural production, and limited opportunities to improve animal production (64). Measures employed to reduce the land area occupied by tsetse have included tsetse control by insecticidal spraying, brush clearing, game elimination, use of

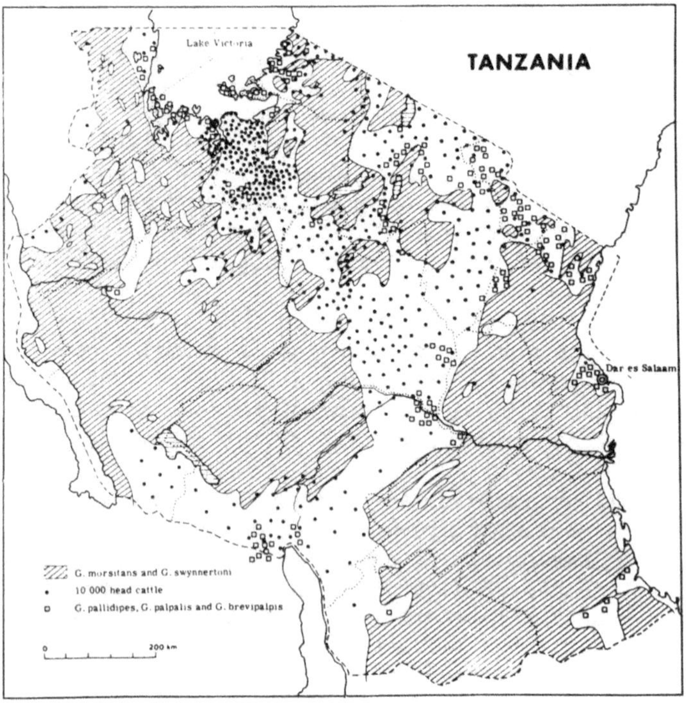

Fig. 6. Distribution of cattle and tsetse fly in Tanzania (63).

trypanocidal drugs, and rearing of tolerant cattle. The use of insecticides to control tsetse has freed an area of about 125,000 square kilometers in northern Nigeria, and the program is continuing at a rate of 12,500 square kilometers per year (3). An extensive program to control trypanosomiasis throughout the entire 7-million-square-kilometer area at a projected cost of $2,250 million over a 40-year period was presented by FAO to the World Food Conference, Rome, November 5-16, 1974. A resolution (no. 11) of the conference called for the FAO to immediately initiate pilot field control projects and applied research in preparation for future large-scale operations to control African animal trypanosomiasis. Studies on the economic and sociological aspects of eradication of tsetse from an area of Uganda and Tanzania (65, 66) have generally supported the view that tsetse eradication and the subsequent utilization of the land for livestock grazing must both be viewed in terms of a comprehensive plan of livestock development and land use. It is critically important to prudently and correctly use the newly-opened land in a manner that will assure its greatest productiveness. The need to protect the environment while still addressing the needs for economic and nutritional benefits is a matter for thorough review (62).

The other major class of arthropods that causes severe economic reductions in the production of livestock, especially cattle, is the ticks (Ixodoidea). There are about 800 species of ticks in the world, and these obligate parasites are reservoirs and vectors of a wide variety of diseases of man and his domestic animals.

Ticks can cause losses in animal production in many ways. Paralysis caused by ticks of the genus *Dermacentor* is limited in America to the northwestern part of the continent (67). Other species of ticks in such genera as *Ixodes*, *Rhipicephalus*, and *Hyalomma* cause paralysis in livestock in Australia, Africa, and Europe (Fig. 7). Heavy infestations of ticks cause intense irritation and general physical damage resulting from licking and scratching by animals. Also, the salivary secretions that ticks inject into animals can cause illnesses such as sweating sickness, caused by *Hyalomma*, and tick toxicosis, caused by *Rhipicephalus* spp., both in southern Africa.

A variety of studies, mostly from Australia and with *Boophilus microplus* (Canestrini), have shown that ticks have many effects on the production of cattle other than disease transmission. When the number of engorging female ticks per day was correlated with the annual reduction in growth rate of Hereford and Friesian heifers (68), an infestation

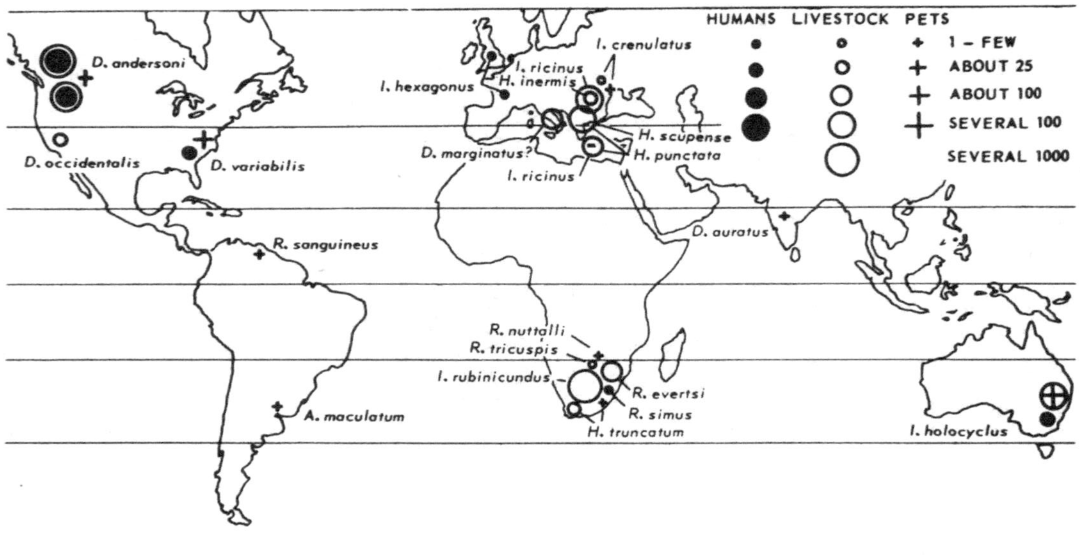

Fig. 7. Distribution of ticks and cases of tick paralysis (67).

of only 50 such ticks caused a reduction of 1.67 pounds per tick, and a burden of 90 engorged females caused a loss of 65 pounds per animal per year in Hereford heifers (69). Nevertheless, when young Hereford or Shorthorn (Bos taurus-type cattle) and Droughtmaster (Bos indicus-type cattle) were both exposed to ticks in a dry, tropical environment, they gained weight at about the same rate though the Bos taurus-type cattle several times had to be treated with an acaricide and had significantly more engorged ticks than the Bos indicus-type (70). The effect of ticks on changes in weights of pregnant and lactating Bos taurus-type and Bos indicus-type crossbred cows in the same area was similar, probably because the Bos taurus-type cows were dipped in an acaricide (71). It was necessary to protect Bos taurus-type steers by dipping them in an acaricide because they had higher tick burdens and gained significantly less weight than the Bos indicus crossbred steers (72). The average gain of 0.28 kilograms per tick per year of dipped cattle compared with undipped cattle, even within a breed, is interpreted to mean that the effect on weight gain is proportional to the number of maturing ticks. However, the number of maturing ticks differs among the breeds of cattle exposed to similar infestations (73).

Other studies (74, 75, 76) with Hereford cattle revealed that changes in hematocrit, hemoglobin, plasma protein-bound iodine, serum globulin, cholesterol, phospholipid, triglicerides, and total protein red cell count were related to tick numbers (Boophilus microplus) and nutrition (high, low, and moderate levels). The separation of anorectic effects (reduced feed intake), which are very different (77) from the specific effects of tick infestation, supported the conclusion that a toxin injected by the ticks interfered with replenishment of albumin, hemoglobin, and cholesterol in protein metabolism (76, 78). In addition, cattle exposed to ticks had a higher respiratory quotient than unexposed cattle though the measurement was taken 4 weeks after cattle were freed of ticks by dipping (79). These data indicate an effect on protein metabolism.

The fact that ticks are vectors and reservoirs of a variety of viral rickettsial, and protozoal diseases of man and his domestic animals makes ticks a major factor in the less than optimum production of animal protein (80). The classic tick-borne diseases of cattle include babesiosis, which is caused by Babesia spp. (Protozoa); anaplasmosis, which is caused by Anaplasma spp. (Rickettsia); theileriosis, which is caused by Theileria spp. (Protozoa); and cowdriosis, which is caused by Cowdria ruminantum (Rickettsia) (81). In addition, a number of other diseases caused by viruses and spirochaets are transmitted by ticks. Usually all these diseases are controlled by controlling ticks.

An example of the magnitude of the problem can be cited. At present there is no practical therapy for East Coast fever, which is caused by Theileria parva (Theiler, 1904) and transmitted principally by Rhipicephalus appendiculatus Neumann. However, this disease, which is found in eastern and southern Africa, has been controlled by frequent (often at 5-day intervals) dipping or spraying of cattle with acaricides to kill the vector before the tick can transmit the disease. (Now, unfortunately, the heavy use of acaricides in certain areas has led to the appearance of strains of ticks that are resistant to many commonly-used acaricides (82).) In Uganda (83), calf mortality is 30 percent in conventional herds in the area where East Coast fever is endemic, but it has declined to less than 5 percent in areas where East Coast fever has been eradicated. The improvement is of great importance because Uganda, while it has overgrazed areas, still has room to expand the cattle population substantially to fully use grazing lands for the production of cattle for needed animal protein. Unfortunately, early importations of Bos taurus cattle usually died within months, so the local adapted breeds such as Zebu, Aukole, and Nyanda were selected for improvement through selective breeding. Such improvement is very slow though it can be speeded up by crossing with exotic beef and dairy cattle. But before the production of high-producing beef and dairy cattle can be increased, East Coast fever must be eradicated. A tick control project for Uganda (83) projected that the eradication of East Coast fever would allow the cattle industry to expand and would thus create an increase in farm income from milk and meat from 20.9 million shillings (2.9 M $) in 1969-70 to 102.2 million shillings (14.3 M $) in 1980-81. This increase could be accomplished on a total budget of 57.3 million shillings (8.0 M $).

As noted, Boophilus microplus is a major concern to cattle raisers in most of northern Australia (Fig. 8). A study of the economics of the involvement of the government of New South Wales in control or eradication of B. microplus, the vector of Babesia spp., concluded that costs of any program for tick eradication (eradication was superior to control) should be paid by all the cattle owners within the area of potential tick distribution (84). Also, the National Government of Australia (85) concluded that although cattle ticks cost governments and producers close to 42 M Australian dollars per year (= $62 M U.S.), eradication of the tick nationally was not practical, but an eradication program in New South Wales was appropriate. The report stressed the need for more use of resistant cattle.

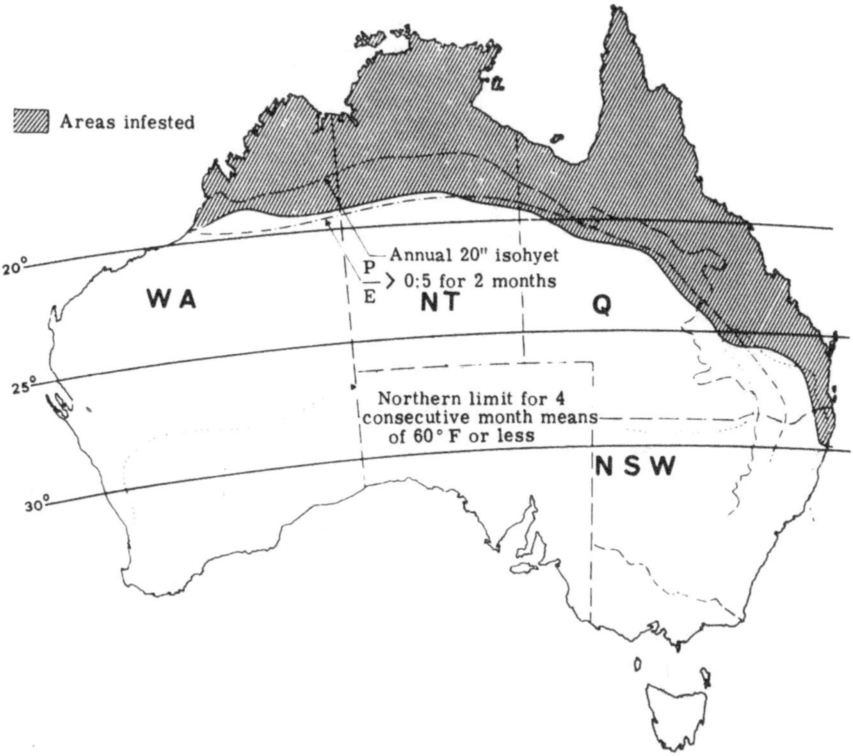

Fig. 8. Distribution of Boophilus microplus in Australia (77).

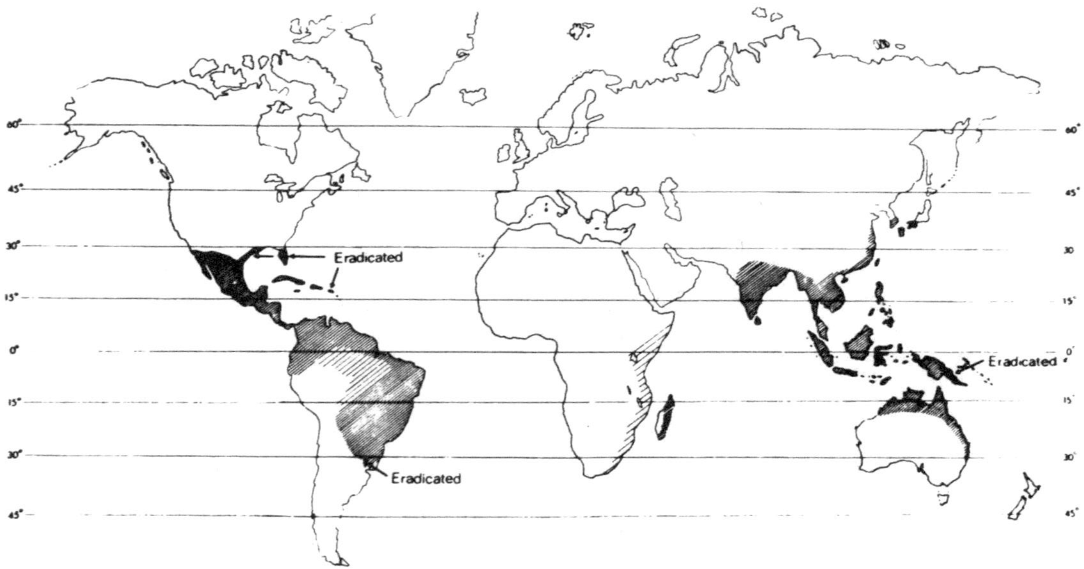

Fig. 9. Distribution of Boophilus microplus in the world (80).

Ticks that affect livestock have been controlled by the application of acaricides that have evolved from crude oil to arsenic to organochlorines to organophosphates and other chemicals. In fact, Boophilus ticks were eradicated from the United States by systematic dipping of millions of cattle in arsenic plus the vacation of pastures, which meant that the larvae died before finding suitable hosts.

Any eradication of ticks is an exceptional accomplishment as is well illustrated by the distribution of Boophilus microplus and the very limited populations of this species that have been eradicated (Fig. 9). A more standard, long-used approach to the cattle tick problem throughout the world (there are exceptions) is that demonstrated by the techniques and philosophies of northern Australia (New South Wales is an exception) where cattle are dipped constantly in acaricides to reduce the number of ticks. A few ticks are left in order to transmit babesiosis to the calves while they are still nursing on the dams. This procedure allows for natural premonition and consequent immunity to the disease. Unfortunately, the constant use of acaricides has put high selection pressure on cattle ticks. The result has been an explosion of resistance (82). The mounting fears that acaricides will not continue to provide the solution to the cattle tick problem in Australia have triggered research activity to provide other solutions.

Alternative measures currently under investigation include pasture rotation, development of resistant strains or breeds of cattle, judicious use of acaricides, and a vaccine to produce immunity to babesiosis. The availability of a vaccine (86) makes possible additional thought of eradicating cattle ticks, but eradication would produce thousands of susceptible animals that have to be protected if infected ticks should reappear in a previously eradicated area. On the other hand, there has been considerable progress in the development of strains of both beef and dairy cattle that have the characteristics of both high production and resistance to ticks. Examples are the Australian Milking Zebu and the Australia Friesian Sahiwal (87). Cross breeding Bos indicus-type cattle with Bos taurus-type cattle rapidly produces tick-resistant cattle; similar selection and crossing for tick resistance within Bos taurus-type breeds are lower to produce highly resistant animals. However, both avenues can develop cattle that are resistant to ticks (88). Also, less treatment of cattle with acaricides for tick control should decrease the effects of these pesticides on the environment.

Mites, another group of arthropods that includes mange and scab mites, affect the production and well-being of cattle (89, 90). Also, the egg production and weight gains of laying hens were significantly adversely affected by the presence of infestations of northern fowl mites, *Ornithonyssus sylviarum* (Canestrini and Fanzago) (91), and of chicken body lice, *Menacanthus stramineus* (Nitzsch) (92). These ectoparasites can be controlled by directly spraying poultry with any of several insecticides.

The total number of livestock in the world has increased in response to the demand for animal products (Table 5). This increase has been dramatic in the case of poultry, probably because of their efficiency in converting plant products into animal proteins. And there are reasonable projections that this increase will continue. However, an increase in numbers of livestock does not assure an increase in the productivity of these animals nor in the equitable distribution of the proteins from these animals among the world's human population. In fact, despite these projected increases in animal numbers, there are projected deficits in the amount of beef produced related to increased needs (Table 6).

Table 5. Trends in world livestock production (93, 94).

Animal	Millions in indicated year 1960	1975	Change (percent)
Cattle*	849.3	1175.0	+ 38.3
Poultry	2647.0	6073.6	+ 129.4
Buffalos	93.8	124.0	+ 32.2
Camels	10.2	14.9	+ 46.1
Goats	316.3	394.8	+ 24.8
Horses	65.4	65.3	- 0.1
Mules	47.9	57.4	+ 19.8
Sheep	1735.3	1060.2	- 38.9
Pigs	483.8	653.9	+ 35.1
Total	6249.0	9619.1	

* Both beef and dairy cattle.

Table 6. Projections of deficits and surpluses of production and consumption of beef (95).

	Deficit of surplus (1000 metric tons) in indicated country or region					
Year	Europe	Latin America	United States	USSR	Other	Total 50 Countries
1965	-545.0	788.1	- 310.8	- 44.9	848.6	736.0
1975	-529.9	1181.1	-1210.8	-122.7	694.7	- 18.3
1980	-550.6	1311.3	-1427.7	-143.4	773.6	- 36.8
1985	-601.7	1428.0	-1689.2	-166.4	857.9	- 171.4

These deficits are greatest in the developed countries. In addition, because of the recognized constraints of energy and land on production of food proteins, additional production must be the result of more efficient use of land and energy (96). The increasing human needs for the plant proteins that presently are fed to livestock demand more efficient production of animal proteins. However a number of major problems must be solved if production of animal proteins is to meet the needs of a growing human population and if the natural resources available for livestock use are to be fully utilized. These problems should be attacked by innovative and carefully planned programs after the development of long-range plans based on each nation's total agricultural resources. Such programs should include determining and using the optimum genetic stock for local conditions, increased forage production and range management, improved livestock nutrition and use of local feedstuffs, especially wastes and byproducts, improved animal husbandry, and the elimination of diseases and pests (10). Limitations on efficiency of animal production include four well-defined areas (97): reproductive efficiency, survival efficiency, efficiency in maintenance of health, and efficiency of conversion of animal feedstuffs to human food.

In the United States the major epizootic diseases of livestock that threaten our intensive livestock production have been eradicated, are under control, or are being excluded by quarantine. The remaining diseases and pests are less dramatic in their effects on production--for example, losses due to helminths and arthropods, stresses from parasites, etc. To attack and solve these problems is the challenge of the future. This challenge has been generally met by applying pesticides to livestock for the control of pests. Obviously, because of environmental considerations, resistance, health hazards, and costs (of production and use), the number of pesticides on the market will be reduced, and the number of new ones developed will be severely limited. Today we seek a better way. Therefore, alternate technologies must be developed that will allow the producer to control these costly pests. However, we still need to prevent losses in livestock due to arthropod pests by developing new pesticides and the technology for their efficient and safe use. The use of these pesticides can then be part of integrated pest management systems that include livestock that are resistant to pests, biological control agents against pests, and innovative technology designed to regulate behavior, development, and reproduction of pests (98).

To accomplish these goals, which realistically include goals in all of the factors that affect the efficiency of the production of livestock, especially ruminant livestock, will take the effort of thousands of well-supported scientists. It has been estimated that world production of ruminant meat and milk could be increased by 50% without increasing land needs or increasing animal numbers simply by placing more research effort on efficiency of animal production (99). Unfortunately, funding for research in agriculture has not kept pace with inflation or the increased needs for additional knowledge to increase production. This situation needs to be rapidly and decisively addressed and changed if there are to be effective solutions to the problems, including those caused by arthropod pests, that limit the production of animal proteins.

REFERENCES

1. N. S. Scrimshaw and V. R. Young, Sci. Am. 235, 51 (1976).
2. D. M. Hegsted, J. Am. Diet. Assoc. 66, 13 (1975).
3. H. A. Jasioroski, in Beef Cattle Production in Developing Countries, A. J. Smith, Ed. (University of Edinburgh, Centre for Tropical Veterinary Medicine, Edinburgh, 1976), p. 2.
4. J. Mayer, Sci. Am. 235, 40 (1976).
5. J. Janick, C. H. Noller, C. L. Rhykeid, Sci. Am. 235, 75 (1976).
6. U. S. Department of Agriculture, Economic Research Service, The World Food Situation and Prospects to 1985 (Foreign Agric. Econ. Rep. 98, Washington, D. C., 1974).
7. W. Lockeretz, Agricultural Resources Consumed in Beef Production (CBNS-AE-3, Center for the Biology of Natural Systems, Washington University, St. Louis, Mo., 1975).
8. R. Revelle, Sci. Am. 235, 165 (1976).
9. E. P. Odum, Fundamentals of Ecology (W. B. Saunders Co., Philadelphia, 1954).
10. W. R. Pritchard, F. N. Andrews, R. F. Dasman, N. L. Garlick, R. J. Kleberg, Jr., N. M. Konnerup, J. A. Pino, C. E. Terrill, in The World Food Problem, Vol. II. Report of the Panel on the World Food Supply (Government Printing Office, Washington, D. C., 1967), chap. 4, p. 241.
11. W. R. Prichard, Proc. Nat. Acad. Sci. 56, 360 (1966).
12. M. M. Kaplan, Social Effects of Animal Diseases in Developing Countries (University of Minnesota Int. Agric. Ser. 3, St. Paul, 1966).
13. P. R. Ellis and M. E. Hugh-Jones, in Beef Cattle Production in Developing Countries, A. J. Smith, Ed. (University of Edinburgh, Centre for Tropical Veterinary Medicine, 1976), p. 105.

14. R. B. Griffiths, in Beef Cattle Production in Developing Countries, A. J. Smith, Ed. (University of Edinburgh, Centre for Tropical Veterinary Medicine, 1976), p. 43.
15. T. E. Gibson, Span 7, 99 (1964).
16. M. H. French, Trop. Anim. Health Prod. 2, 1 (1970).
17. E. E. Miller, J. Trop. Geogr. 41, 59 (1975).
18. L. E. A. Symons, Int. Rev. Trop. Med. 3, 49 (1969).
19. M. M. H. Sewell, in Beef Cattle Production in Developing Countries, A. J. Smith, Ed. (University of Edinburgh, Centre for Tropical Veterinary Medicine, 1976), p. 138.
20. I. A. Barger, W. H. Southcott, V. J. Williams, Aust. J. Exp. Agric. Anim. Husb. 13, 351 (1973).
21. F. B. Roseby, Aust. Vet. J. 46, 361 (1970).
22. H. McL. Gordon, Proc. Aust. Soc. Anim. Prod. 10, 180 (1974).
23. A. D. Donald, ibid. 10, 148 (1974).
24. R. Winks, Aust. Vet. J. 46, 8 (1970).
25. D. L. Ferguson, D. A. Reynolds, M. J. Twiehaus, Can. J. Comp. Med. 35, 82 (1971).
26. R. K. Keith, Aust. Vet. J. 44, 326 (1968).
27. L. Michaud, Can. Vet. J. 8, 85 (1967).
28. A. D. Donald, Victorian Vet. Proc. 1968-69, 34 (1969).
29. I. K. Hotson, Aust. Vet. J. 39, 108 (1963).
30. W. W. Becklund, Am. J. Vet. Res. 25, 1380 (1964).
31. C. D. Steelman, Annu. Rev. Entomol. 21, 155 (1976).
32. A. L. Everett, I. H. Roberts, J. Naghski, J. Am. Leather Chem. Assoc. 66, 118 (1971).
33. W. A. Nelson and S. B. Slen, Exp. Parasitol. 22, 223 (1968).
34. R. H. Wharton, Changing Patterns in Entomology, 18 (1974).
35. G. J. Shanahan and N. A. Roxburgh, J. Aust. Inst. Agric. Sci. (December), 249 (1974a).
36. ______________________, Pest Artic. News Summ. 20, 190 (1974b).
37. M. J. Whitten, G. G. Foster , W. G. Vogt, R. L. Kitching, T. L. Woodburn and C. Konovalov, Proc. XV Int. Cong. Entomol. 129 (1976).
38. J. A. Shemanchuk, W. O. Haufe, C. O. M. Thompson, Can. J. Comp. Med. 24, 158 (1960).
39. R. C. Collins and L. W. Dewhirst, J. Am. Vet. Med. Assoc. 146, 129 (1965).
40. W. A. Nelson, J. A. Shemanchuk, W. O. Haufe, Exp. Parasitol. 28, 263 (1970).
41. K. B. W. Utech, R. H. Wharton, L. A. Wooderson, Aust. Vet. J. 45, 414 (1969).
42. D. G. Ely and T. L. Harvey, J. Econ. Entomol. 62, 341 (1969).
43. F. A. Thrift and J. R. Overfield, J. Anim. Sci. 38, 1179 (1974).

44. Anonymous, Farmer-Stockman 89, 30 (1976).
45. R. R. Blume, J. J. Matter, J. L. Eschle, Environ. Entomol. 2, 811 (1973).
46. K. Romaniuk, Acta Parasitol. Pol. 21, 413 (1973).
47. ________________, ibid. 22, 29 (1974).
48. W. N. Beesley, Vet. Med. Rev. 4, 334 (1974).
49. G. B. Rich, Can. J. Anim. Sci. 50, 301 (1970).
50. H. Thornberry, Ir. Vet. J. 30, 83 (1976).
51. R. C. Bushland, Bull. Entomol. Soc. Am. 21, 23 (1975).
52. U. S. Department of Agriculture, Animal and Plant Health Inspection Service, Progress in Screwworm Eradication (USDA-APHIS 91-25, Washington, D. C., 1974).
53. M. W. Service, Biol. Conserv. 3, 113 (1971).
54. E. Duren and L. E. O'Keeffe, Anim. Nutr. Health 27, 3 (1972).
55. J. B. Campbell, J. Econ. Entomol. 69, 711 (1976).
56. J. L. Eschle, J. A. Miller, C. D. Schmidt, Nature (London) 265, 325 (1977).
57. R. W. Miller, L. G. Pickens, N. O. Morgan, R. W. Thimijan, R. L. Wilson, J. Econ. Entomol. 66, 711 (1973).
58. J. Monty, Rev. Agric. Sucr. Ile Maurice, 51, 13 (1972).
59. C. D. Steelman, T. W. White, P. E. Schilling, J. Econ. Entomol. 65, 462 (1972).
60. ________________, ibid. 66, 1081 (1973).
61. ________________, ibid. 69, 499 (1976).
62. W. E. Ormerod, Science 191, 815 (1976).
63. P. Finelle, World Anim. Rev. 10, 15 (1974).
64. J. T. Snelson, FAO Plant Prot. Bull. 23, 103 (1975).
65. C. E. Bell, Jr., R. O. Gilden, R. O. Wheeler, Tsetse Fly Eradication and Livestock Development (U. S. Department of Agriculture-U.S. Department of State, Agency for International Development, Washington, D. C., 1969).
66. H. E. Jahnke, Forschungsberichte der Afrika-Studienstelle 48 (1974).
67. J. D. Gregson, Can. Dep. Agric., Res. Branch, Monogr. 9 (1973).
68. D. A. Little, Aust. Vet. J. 39, 6 (1963).
69. J. Francis, Proc. Aust. Soc. Anim. Prod. 3, 130 (1960).
70. L. A. Y. Johnston and K. P. Haydock, Aust. Vet. J. 45, 175 (1969).
71. ________________, ibid. 47, 295 (1971).
72. R. W. Gee, M. H. Bainbridge, J. Y. Haslam, ibid. 47, 257 (1971).
73. H. G. Turner and A. J. Short, Aust. J. Agric. Res. 23, 177 (1972).
74. J. C. O'Kelly and G. W. Seifert, Aust. J. Biol. Sci. 22, 1497 (1969).
75. ________________, ibid. 23, 681 (1970).
76. J. C. O'Kelly, R. M. Seebeck, P. H. Springell, ibid. 24, 381 (1971).

77. P. H. Springell, World Anim. Rev. 10, 19 (1974).
78. P. H. Springell, J. C. O'Kelly, R. M. Seebeck, Aust. J. Biol. Sci. 24, 1033 (1971).
79. J. E. Vercoe and J. C. O'Kelly, Proc. Aust. Soc. Anim. Prod. 9, 356 (1972).
80. R. A. Bram, World Anim. Rev. 16, 1 (1975).
81. G. Uilenberg, ibid. 17, 8 (1976).
82. R. H. Wharton, in Control of Arthropods of Medical and Veterinary Importance, R. Pal and R. H. Wharton, Eds. (Plenum Press, New York, 1974), p. 35.
83. D. S. Ferguson and T. T. Poleman, Modernizing African Animal Production: The Uganda Tick Control Project (Cornell Int. Agric. Mimeogr. 42, Cornell University, Ithaca, 1973).
84. J. H. Johnston, Rev. Marketing Agric. Econ. 43, 3 (1975).
85. Commonwealth of Australia, Cattle Tick Control Commission, Cattle Tick in Australia -- Inquiry by the Cattle Tick Control Commission (Australian Government Publishing Service, Canberra, 1975).
86. L. L. Callow, World Anim. Rev. 18, 9 (1976).
87. I. Byford, P. Colditz, R. Sibbick, Queensl. Agric. J. (January-February), 11 (1976).
88. R. H. Wharton, Anim. Quart. 4, 13 (1975).
89. I. H. Roberts, Great Plains Beef Cow-Calf Handbook, Coop. Ext. Serv. GPE-3255, 3255.1 (1975).
90. P. T. Diplock and R. H. J. Hyne, N.S.W. Vet. Proc. 75, 31 (1975).
91. J. G. Matthysse, C. J. Jones, A. Purnasiri, Search Agric. 4, 1 (1974).
92. J. A. DeVaney, Poult. Sci. 55, 430 (1976).
93. Food and Agriculture Organization of the United Nations, Animal Health Yearbook (1961).
94. Food and Agriculture Organization of the United Nations, Animal Production and Health Division, Animal Health Service, Animal Health Yearbook, H. O. Konigshofer, Ed. (1976).
95. J. R. Simpson and D. E. Farris, Texas Agric. Expt. Stat. Consolidated PR-3383, 110 (1976).
96. D. Pimentel, W. Dritschilo, J. Krummel, J. Kutzman, Science 190, 754 (1975).
97. National Academy of Sciences, Agricultural Production Efficiency (Washington, D. C., 1975).
98. ______________________, World Food and Nutrition Study (Washington, D. C., 1975).
99. T. C. Byerly, Science 195, 450 (1977).

6

Post-Harvest Food Losses: The Need For Reliable Data

John R. Pedersen

Introduction

Over the past several years, considerable emphasis has been placed on increasing production of the world's food supply to meet the needs of a growing population. High yielding varieties of rice and wheat have been developed and in many areas of the world there have been material increases in the available food supply.

Only recently has there been a realization, on a broad scale, that we need to increase our efforts in the area of post-harvest food loss reduction as a means of increasing the total available food supply. This is evidenced by the emphasis placed on post-harvest losses at the World Food Conference held in Rome, 1974; recent programs undertaken by FAO, The National Academy of Sciences, The U.S. Agency for International Development; and by many other national and international organizations.

Since the late 1940's, we have seen a variety of loss estimates published for various parts of the world. These estimates have ranged from very low percentages to as high as 50 percent and greater for stored food losses in the developed and developing countries. Post-harvest food losses are attributed to a variety of causes. In the durable commodities, such as cereal grains, legumes, pulses, oilseeds, etc. losses are due primarily to insects, rodents and fungi. In the perishable commodities, such as vegetables, fruits, meats, etc., losses are due primarily to microbial spoilage.

The food sector where most significant post-harvest food losses are probably occurring is in the durable commodities (cereal grains, etc.). Most studies that are available primarily refer to the durable commodities and it is the food sector on which this discussion is based.

Personal observations, discussions and literature make it apparent to me that the post-harvest food loss figures or percentages that have been published and quoted, in many cases may be high and in most cases are not backed up by "hard data". That is, many of the estimates of post-harvest losses that we see and hear are estimates based on judgements and, in many cases, guesses as to what losses are actually occurring.

Recently, the Tropical Products Institute (London, England) assumed the responsibility, within GASGA (Group for the Assistance on the Storage Grains in Africa), to coordinate the work on loss assessment in durable produce (cereal grains, legumes, etc.). As part of its overall undertaking, TPI has started to evaluate the existing literature on post-harvest losses in durable produce. As of June of 1976, 93 records of losses had been classified as "guesstimates", 91 as supported estimates (primarily laboratory-type records) and 9 reliable estimates (1,3). Most of the reliable estimates are for losses caused by insects, in cereal grains stored at the farm level in Africa. It is obvious that we do not have a depth of reliable loss assessment data.

Food Losses Occur At All Levels In The Post-Harvest System

Losses occur in the field, farm storage, village storage, central storage, in-transit, processing, marketing and at the point of consumption.

Field Losses

Losses which occur in the field may be debated as being pre- or post-harvest losses. There are losses, however, that do occur to crops, such as maize, that are left on the stalks for storage in the field. Typical storage insect pests, rodents and birds are known to cause losses to "field stored" grain. Even termites, insects not generally considered "food pests", cause considerable loss in field stored maize by weakening the stalks causing them to lodge and then consuming the ears of maize.

Farm Losses

For the developing countries, estimates indicate that 70 to 90 percent of the cereal grains produced are stored and consumed on the farm. Since a considerable amount of the production is stored on the farm in some rather primitive type structures, it might be assumed that the greatest losses are occurring on the farm. This may or may not be true.

Personal observations and the literature indicate that farmers in several of the African countries take relatively good care of the grains they intend to consume themselves or use for seeds. This occurs through use of indigenous storage methods and probably a selected resistance of indigenous varieties of grains.

As an example, maize for home consumption is commonly stored in the husk over cooking areas in some African countries (11,14). Heat from cooking fires probably has a drying effect and smoke probably makes the situation less acceptable for insect development. On the other hand, maize to be marketed is commonly left in the field until sold. Here it is exposed to the elements and infested from crop residues, termites, rodents and birds.

Village-Local Dealer Storage

Grains sold to dealers and held in small village stores might be expected to be of lower quality than that retained on the farm. In the developing countries, farmers tend to save the better grains for their storage. This is practiced in the developed countries as well.

For this reason, grain in local dealer storage is likely to show greater levels of damage and loss. Local dealers are less likely to hold grains for extended periods of time and probably are not concerned with losses.

Information on losses that occur in local dealer storage is virtually non-existent.

Central Storage

Central storage in the developing countries is commonly operated by the government. At this point in the post-harvest system, it should be easiest to develop reliable data on grain condition and losses. However, only a few studies at this level have been reported. Probably the best study at this level was conducted on groundnut storage in Nigeria in 1948-49 (9).

Records commonly kept at central storage facilities, i.e. in and out weights, inspection and treatment records, could, with a small additional effort, be used to generate reliable loss estimates at this level in the post-harvest system.

In-Transit Losses

Reliable estimates of in-country transit losses are non-existent for developing countries.

At present, the U.S. Department of Agriculture is conducting a study to determine the changes in quality of grains during transport in export shipments. An earlier study (8), included only physical grade factors, however, the current study includes an infestation evaluation also. The U.S.S.R. and other countries have also conducted recent studies on in-transit overseas shipments.

Processing

Very few studies have been reported where losses have been quantified for processing operations in developing countries. Losses can occur in home processing, small custom processing or large scale processing. It is more likely that lower quality may be a more significant factor in processing than physical loss of commodity.

Marketing

Losses occur in the marketing of grains and processed cereal products. Materials held for any length of time in wholesale or retail markets undoubtedly suffer losses but to my knowledge, no studies of this type have been conducted in the developing countries.

Consumer

Wastage of foods at the consumer level is probably responsible for losses that do occur. The extent of loss or degree of wastage is probably closely correlated with the degree of food abundance. Food not consumed by humans is often times consumed by animals and thus some value is obtained.

Measurement Of Post-Harvest Food Losses

Two areas of concern are encountered in measuring post-harvest food losses: defining what constitutes a loss and methods for measuring the defined loss.

Methods Of Expressing Losses

One of the first things that needs to be resolved in making a loss assessment is--what constitutes a loss? There are a variety of types of losses which have been considered--weight loss; nutritional and energy loss; quality loss; monetary loss; and others such as seed loss, loss of goodwill, etc. (7).

Weight loss is considered the easiest type of loss measurement to make and is the measure most commonly used in arriving at most of the estimates that have been published. There are problems in using total reduction in weight in expressing losses, however. Loss or gain in moisture content of commodities, such as the cereal grains, can either exaggerate or mask a loss. For this reason, dry matter weight loss--referred to by some as "food loss"--is considered the most reliable loss expression. Weight losses can be expressed in real terms such as tons, bushels, etc. but are more commonly presented as percent lost.

Nutritional loss and energy loss provide measures of the loss of food value of commodities to humans and/or animals. We can measure reduction in protein, carbohydrates, fats, vitamins, etc. but these values are somewhat difficult to express as losses readily visualized as reduction in the total food supply. Analyses required to measure this type of loss are complex, costly and not as readily adapted to application in the developing countries. Recent work in Canada has been directed at developing a workable criterion on energy loss (in calories) caused by populations of specific insects on certain cereal grains. It is hoped that, through this work, a loss measurement criteria might be developed that would be universally acceptable for budgeting food resources, and developing rational storage and distribution systems for food grains (13).

Quality loss as a measurement of post-harvest food loss has considerable merit, however, as a general measurement of food loss there is a serious problem in standardizing what level of quality is considered sub-standard and actually constitutes a loss. Not only are there difficulties in defining what constitutes a quality loss but there are different qualities acceptable in different countries. As an example, one insect in a 5, 10 or even in a 100 pound bag of flour is considered unacceptable here in the United States causing the product to be either discarded or down-graded to animal feed, whereas in other parts of the world several insects in a commodity might be considered inevitable. Definitions of quality can change based on the lack or availability of a food also. In Zambia, for example, infested ears of maize are more apt to be fed to animals than used for human consumption immediately after harvest when maize is abundant. Going into the more difficult storage period of the year, more and more of the infested grain will be consumed as maize becomes limited in availability (2). Although quality loss really tells us what may be acceptable or not acceptable as food, it is a hard form of loss to quantify.

Monetary loss is one way that other forms of loss can be expressed. It puts a dollar value on the loss and can allow us to make economic comparisons with alternatives to the losses. In some cases, the cost of applying fumigants, insecticides and other forms of pest control may be considered a monetary loss. There are some of us who feel that these costs should be considered a maintenance cost or expense much like changing oil in a car or maintaining the condition of a home. Personal observation in developing countries has indicated that some of the persons in charge of food stocks are not willing to accept the fact that monies expended to maintain these stocks is a business expense and not a loss. This is probably true in the developed countries of the world also.

Methods For Measuring Losses

Estimates of post-harvest food losses are obtained in a variety of ways--judgements, survey questionnaires, laboratory studies, field spot sampling and detailed sampling studies.

Judgements are most often based on observations or impressions and generally not substantiated by measurements. The reliability of these estimates is affected by the experience of the observer and the extent of the observations. Most of the loss estimates we see published or quoted in the various media today, ranging from 0 to over 50 percent of foods lost post-harvest, are based on this type of estimate. More accurately, these figures should probably be stated as "guesstimates." There are indications also that percent loss values that have been profusely published are, in many cases, overstated (10). I believe we tend to remember the more dramatic or catastrophic things we observe and as a result, may tend to apply these observations to a total situation.

Survey questionnaires have been used quite extensively in arriving at loss estimates, however, they are subject to a series of biases, even if well designed. Interviewees often give inaccurate responses, either innocently or in an attempt to deceive the interviewer. Surveys, especially those conducted at the farm level, may yield inaccurate responses because of fear of some type of government reprisal or because the interviewee is generally apprehensive. A thorough loss study conducted in Zambia indicated that survey questionnaires yielded fairly reliable data when taken near the main loss project where farmers were familiar with the work. Questionnaires completed where the project was unknown were considerably less reliable (4). In 1968, a survey of the Asian Productivity Organization member countries reported that from 5 to 10 percent of all food grains produced were lost during storage and distribution (APO Project No. SUV/I/68). When several neighboring countries share information of this type, there may be a tendency to "err" on the low side in the interest of national pride. It is interesting to note in this case, that none of the APO member countries reported losses exceeding the widely published FAO estimate of 10 percent world-wide losses (7). There are indications also that biases may occur within countries when comparing government versus non-government storage losses.

Laboratory studies on losses caused by the various factors of deterioration (insects, rodents, birds, fungi, etc.) help us to develop techniques for measuring losses and provide insights into what might be expected to occur in the field. However, very few, if any, loss estimates developed in the laboratory could be extrapolated to actual field conditions.

Field spot sampling is often used in conjunction with survey questionnaires or other surveys to assess what losses are occurring. One hazard to this form of loss assessment is that a loss at only one point in time may be observed. If the observation is made shortly after harvest, a low level of loss probably will be found. If the observation is made late in the storage year, a greater loss will probably be observed. If an annual loss estimate of total production is based on the early observation, the estimate will be low. If the annual loss estimate of total production is based on the observation made late in the storage period, the estimate will be high. The problem lies in making an accurate estimate in a situation where there is a decreasing quantity of food and potentially, an increasing degree of loss. A hypothetical case shown in Figure 1 is a very close approximation of an actual study conducted in Zambia (4). The degree of loss observed near the end of the storage period applied to total production would indicate a loss of 20 percent. Considering a decreasing supply and loss based on a series of observations taken throughout the storage period would indicate a loss of roughly 13.5 percent. Very few references to losses in the literature indicate whether this is considered in making the estimate.

Detailed loss studies under field conditions have been the exception rather than the rule. Earlier it was indicated

Figure I. Hypothetical Calculation of Grain Loss

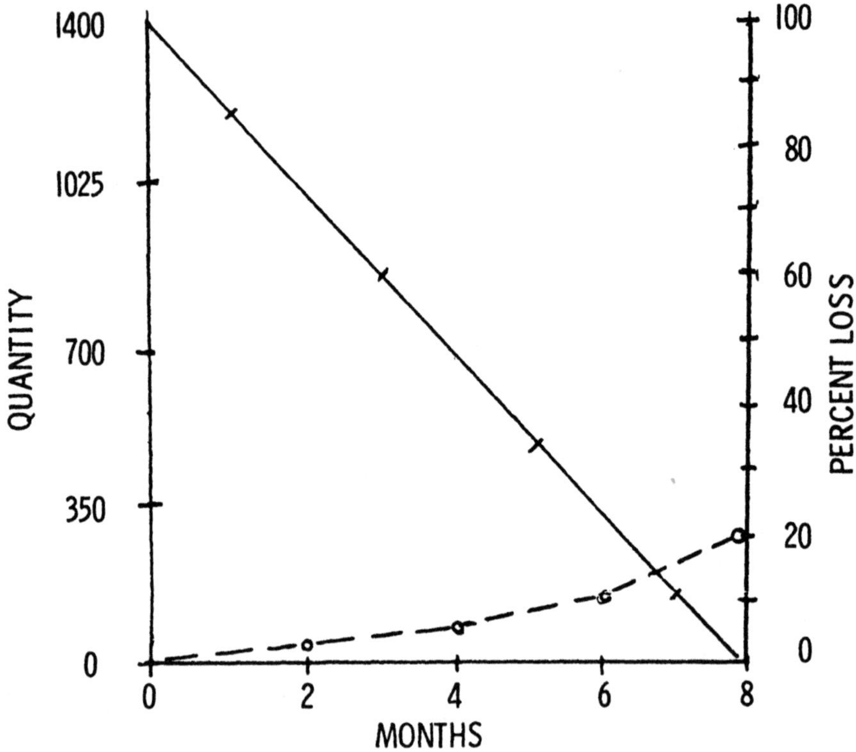

1400 pounds at 20% loss = 280 pounds

Graduated loss:

2 mos.	1225 pounds at	3%	36.75	pounds
4 mos.	875 "	6%	52.50	"
6 mos.	525 "	12.5%	65.63	"
8 mos.	175 "	20.0%	35.00	"
	Graduated loss		189.88	"
	Graduated % loss		13.5%	

that TPI, in its literature search on durable commodity losses, classified approximately nine references out of close to 200 as providing a thorough analysis of losses under field conditions. The difficulty of arriving at reliable estimates of losses varies considerably under given situations. In countries such as Australia, where the largest portions of grains marketed are controlled by the government and good records are kept, it is possible to make accurate estimates of losses that occur (5). Here yearly estimates of 0.14 to 0.68% have been reported over a 10 year period. In the United States where grain is traded on the open market, it would be quite difficult to make an accurate assessment of what level of loss actually occurs. I know of few published studies of losses in U.S. commercial storage. At the farm level, very few thorough loss studies have been conducted either in the developed or developing countries. Recent studies in Malawi (12) and Zambia (4) provide guidelines for future studies. In the Zambia study, a complete analysis of the storage environment, losses, quality criteria and economics were considered. Obtaining reliable loss estimates is an expensive and time consuming effort. This is probably one of the reasons accurate loss assessment hasn't received the emphasis it should in the past.

Use Of Post-Harvest Loss Data

Do we need accurate estimates of losses? Perhaps it depends on what we want to do with the estimates. If we want to "shake things up" and get programs moving, somewhat shocking (although unreliable) estimates might provide the stimulus. Some might refer to this as "political expediency." If this were true, a considerable amount of action should have occurred before now. We have had estimates of 30 to 50 percent loss for certain of the tropical developing countries since the late 1940's and a 10 percent world-wide loss estimated by FAO (7).

The concern for doing something about post-harvest food losses was well expressed in a paper on post-harvest technology prepared in 1974 for presentation to the Technical Advisory Committee of the Consultative Group on International Agricultural Research. It stated, "the most conservative estimates suggest annual post-harvest grain losses of the order of millions of tons. Whether viewed in economic terms or in terms of human nutrition in the LDC's, such losses are intolerable. Consequently, the principal gain from more efficient post-harvest systems would be a substantial increase in the food gains available in the LDC's." The paper further recommended that "potential benefits justify a considerably increased investment in post-harvest research, development, information and training."

Other research and technical assistance organizations are also increasing their efforts in the area of post-harvest food loss reduction. FAO appears to have accepted the challenge presented in the September 19, 1975 resolution of the Seventh Special Session of the United Nations General Assembly which stated that "the further reduction of post-harvest food losses in developing countries should be undertaken as a matter of priority, with a view to reaching at least 50 percent reduction by 1985." The U.S. Agency for International Development has also designated post-harvest loss reduction as a Key Problem Area and is currently funding at least two projects reviewing the extent of post-harvest food losses and development of loss assessment methodology as a basis for determining where the greatest impact can be made with available resources.

It is obvious, with the increased emphasis being placed on post-harvest technology and food loss reduction, that considerable amounts of capital and time will be devoted to reducing the losses. Reliable post-harvest loss data will be a necessity in allocating these resources where they will provide the greatest benefit. A few examples will serve to indicate situations where reliable loss data are needed.

We need to know what losses are occurring at various points in the post-harvest system. It is commonly assumed that greatest losses occur at the farm level with a general tendency for the degree of loss to decrease as the cereal grains proceed through the post-harvest system (15). There is no assurance that this is true in every country nor in different areas of a given country. Before an effective loss reduction program can be implemented within a country, it is necessary to know what and where losses are occurring and an analysis made as to whether the losses can be economically reduced. This requires reliable loss data.

Loss estimates and projected loss reductions were used as a basis for justifying over half of a $73.3 million proposed storage, handling and processing facilities project for Indonesia (16). In a situation such as this, where the investment of large sums of capital are based on anticipated reduction of losses, the loss estimates must be reliable in order to develop reliable benefit/cost ratios.

It has been suggested that rice be given priority for consideration in loss reduction efforts because (1) on a world-wide basis rice is the second most important crop after wheat, (2) it is the most important cereal crop for the developing countries, (3) rice is grown in the majority of the developing countries, (4) it is grown under tropical conditions where intensive biological activity and humid conditions favor post-harvest losses, and (5) being the staple food of many low-income people, reduction of losses will help the poorest (6). Although the assumptions stated above are sound, very few studies on rice losses have been published and reliable data is lacking.

Interventions or efforts to reduce losses will require the expenditure of capital. Whether or not a farmer, for example, can afford to pay the cost of reducing his losses will have to be determined based on an accurate assessment of his losses and the cost of improved storage. Some indigenous storage techniques are quite effective in minimizing losses. Comparative studies to determine which indigenous storage practices are most efficient should be conducted and the most efficient promoted. Many of the developing countries have very poor or no extension services and to establish such service would be time consuming and costly. Nations may be required to determine whether extension activities are warranted.

It has already been mentioned that donor agencies, such as the U.S. Agency for International Development and others are concerned with the extent of post-harvest food losses. They need guidance in allocating their resources so that the most severe loss problems are attacked first. In order for them to place their resources where they will be most beneficial, reliable loss data is required.

Our ultimate goal is to provide an adequate diet for all peoples of the world. One of the ways we can work toward accomplishing this is by minimizing our post-harvest food losses. Reliable information on what losses are occurring is a necessary benchmark on which we should base our efforts to reduce food losses. If we are to reduce post-harvest food losses 50 percent by 1985, as has been suggested, we should have a "benchmark" on which to base the 50 percent reduction.

References

1. Adams, J. M. (1976) Report on post-harvest loss assessment in durable produce, with particular reference to methodology. Paper presented at the Group for Assistance on Storage of Grains in Africa (GASGA), Montpelier, France 23-25 June 1976, 11 p.

2. Adams, J. M. (1976) Personal discussions.

3. Adams, J. M. (1977) A cross-referenced bibliography on post-harvest losses in cereals and pulses with particular reference to tropical and subtropical countries. Tropical Products Institute, Report G100., 28 p.

4. Adams, J. M. and G. W. Harman (1977) The evaluation of losses in maize stored on a selection of small farms in two areas of Zambia, with particular reference to methodology. Tropical Products Institute Report G109, 149 p.

5. Bourne, Malcolm C. (1976) Promising avenues for increasing efforts in reducing post-harvest food losses in developing countries. A report to the Technical Assistance Bureau, Agency for International Development, June 1976, 23 p.

6. FAO (1975) Reducing post-harvest food losses in developing countries. AGPP: MISC/21, 60 p.

7. Hall, D. W. (1970) Handling and storage of food grains in tropical and subtropical areas. FAO Agricultural Development Paper No. 90, Rome:FAO, 350 p.

8. Hill, L. D., M. R. Paulson and B. L. Brooks (1976) Grain quality losses between origin and destination of export grain -- a case study. AE-4399 Dept. Agr. Econ., Univ. of Illinois, 8 p.

9. Howe, R. W. (1965) Losses caused by insects and mites in stored foods and feeding stuffs. Nutr. Abst. and Rev. 35:285-293.

10. Lipton, M. (1971) Research into the economics of food storage in less developed countries: Prospects for a contribution from U.K. technical assistance. Institute of Development Studies, Univ. of Sussex Comm. Series No. 61, 17 p.

11. Mphuru, A. N., M. A. M. Maro and L. A. Odero-Ogwel (1974) Traditional Storage of Food Grains in Tanzania - with particular reference to the storage of maize in Iringa and Morogoro regions. Univ. of Dar Es Salaam, Faculty of Agriculture and Forestry, August 1974, 57 p.

12. Schulten, G. G. M. (1975) Losses in stored maize in Malawi and work undertaken to prevent them. EPPO Bul. 5(2):113-1110.

13. Sinha, R. N. and A. Campbell (1975) Energy loss in stored grain by pest infestation. Canada Agriculture, Spring 1975.

14. Sorenson, L. O., J. R. Pedersen and N. C. Ives (1975) Maize Marketing in Zaire. Grain Storage, Processing and Marketing Report No. 51, Food and Feed Grain Institute, Kansas State University, Manhattan, Ks, 262 p.

15. Spensley, P. C. (1975) Post-harvest food losses and ways of reducing them. Paper presented at the 2nd Latin-American Congress on Food Science and Technology, Campinas Brazil, August 1975. 23 p.

16. Weitz-Hettelsater Engineers (1972) Economic and Engineering Study. Rice Storage, Handling and Marketing -- The Republic of Indonesia. Prepared for Agency for International Development. 722 p.

7

Of Millet, Mice and Men: Traditional and Invisible Technology Solutions to Post-Harvest Losses in Mali

Hans Guggenheim

Introduction

Inadequate farm-and-village-level storage is often blamed for post-harvest crop losses of 20-40% according to statistics for sub-Saharan Africa (1). Our research on post-harvest damage to millet in traditional Dogon graneries in Mali indicates average losses of only 2-4% annually, but reveals high losses in warehouses operated by official marketing boards.

Because of the low productivity of the average farmer (350-500 kg/ha) and the continuing deficiencies in production, no more than 10-15% of the crop can be commercialized by the official marketing board for distribution to urban centers and adjacent deficiency regions. Losses to pests probably equal this amount.

This paper argues that current official marketing practices are one of the major impediments to increased production, and that a buffer-stock scheme supporting the parallel market would be more appropriate in stimulating production and regulating the millet market than the present system of control. In order to develop such a scheme, modern storage facilities need to be improved, and traditional farm storage integrated into the process. The technical research on losses to pests in traditional and marketing-board storage, and the technical recommendations for ameliorating current conditions are meant to provide a basis for the formulation of alternatives to current practices. Although our primary research has centered upon the Dogon, the major millet farmers of the Fifth Region, and upon the activities of Operation Mils in Mopti, we hope that the conclusions of this paper may have broader applications.

Mali and its Population

Mali covers 1.2 million km^2 in three zones -- 280,000 km^2 of desert (less than 200 mm annual rainfall, 320,000 km^2 in the Sahel (300-600 mm annual rainfall), and a Sudanic zone of 600,000 km^2 (over 600 mm annual rainfall). Mali's current population, as measured by the 1976 census, is 6.3 million. If this is correct, the 1970 census (which estimated a population of 5 million and a 2.5 % annual growth rate) was in error, and current projections of future food needs must be pushed upward.

Ninety-one percent of Mali's population is involved in agriculture, which accounts for 44% of the GNP. The lack of an adequate transportation system poses one of the country's most serious problems and makes fundamentally significant the issue of storage. In 1970, 3.5 million people lived in villages of fewer than 1000 inhabitants, villages often isolated by floods in the rainy season and by sand dunes in the dry season. They are difficult to supply with food in emergencies and difficult to integrate into the economy as grain suppliers in good years. The costs and risks of transport are too high in relation to the price of grain.

Grain constitutes the basic diet of the population. Government objectives are therefore to keep prices low in urban areas, to insure enough grain for deficiency areas, and to develop an export capacity that will support increased production. Such objectives may require policies different from those needed to reduce rural poverty and to reduce the risks farmers perceive in selling their grain reserves as well as in investing in new production methods.

An annual increase of 6% in supplies for Bamako and of 4.5% for towns of 2000-4000 inhabitants will be needed if per capita consumption is to be held at 160 kg a year with a population increase of 2.5 - 3%. To insure such increases will not be easy. Much of Mali's available land is in marginal zones of low and uncertain rainfall and fragile soils. Given current pricing policy, intensive agriculture based on modern technology poses an economic gamble to farmers, and is not necessarily advisable in terms of long-term resource management. Production, pricing, and storage goals of Op Mils must be interpreted in terms of these constraints.

An important report from the Center for Research on Economic Development (CRED) at the University of Michigan calculated that Mali has already achieved close to nutritional self-sufficiency in cereal production. Production for 1976 of coarse cereals was near 975 thousand tons and of rice, 169 thousand tons. CRED figures a deficit of only 152 thousand tons and then suggests that this deficit is illusory, that a disguised surplus may be evidenced by parallel market prices 15-30% below the official market price. CRED's optimism may be premature (2).

The inadequate distribution and storage systems account in part for seasonally-and-regionally-occurring low prices on the parallel market. In fact these prices fluctuate widely, rising in August to double or triple the level of the official market price. The need of small, rural households to sell grain locally "by the spoonful" in order to meet immediate needs and pay taxes is also responsible for some low prices. Neither of these factors is disguising a surplus.

Neither our research in the Fifth Region nor our analysis of Op Mils production statistics gives much support to conclusions that nutritional self-sufficiency has been achieved. According to its director, Op Mils's published statistics for production and yields are frequently inflated, but statistics for commercialized grain appear accurate (3). This is because increases in commercialization reflect more efficient collection methods over increasingly extended areas and in more marginal zones rather than increases in yield averages or production. The illusion of progress created by these statistics can lead to commercialization and storage policies which, in case of drought, add to the risks of the rural population (by reducing local reserves) and to those of the government (which would have to rush emergency supplies into stricken areas).

The Geography

Topographically, the Dogon region is centered around the Bandiagara plateau, which slopes gently from the west to a series of steep escarpments at its eastern edge. Most elevations on the plateau range from 450 to 600 meters; isolated spots are even higher. The average width of the plateau is about 75 kilometers, and its length 200 kilometers. The inland delta of the Niger borders the plateau to the west and north (cercles of Mopti and Douentza), and to the south and east lies the Gondo plain (cercles of Koro and Bankass). The elevation of the Gondo plain at its western edge, where it abuts the cliffs, is between 250 and 300 meters. It slopes east to the Gondo Valley depression.

There is relatively little surface drainage beyond the near neighborhood of the cliffs. Water resources are thus limited to wells. The geological stratification demands that wells often exceed 60 meters in depth; even then it is highly

Table 1 Differences of Production and Yield and Quota Figures for the Circle of Bankass Between 1975/76 and 1976/77

Z.E.R.	Increase in Surface Between 1975/76 and 1976/77	Yield Differences Between 1975/76 and 1976/77	Difference in Production	Difference in Quotas
Centrale	+20 ha	+150	+830	+100
Diallassagou	+50	+50	+337	+150
Koulogan	+30	+250	+2067	+250
Baye	+16	+150	+830	+100
Kani-Bonzon	-	-160	-698	-
Ségué	+200	+280	+874	-
Sokoura	+5	-100	-668	-40
Ouenkoro	+100	+230	+648	+200

Table 2 Yield, Production and Quota at Bankass for 1975/76 - 1976/77

Bankass

Z.E.R.	MILS									
	Surface (Hectare)		Yield Kg/ha		Production in Tons		Quota in Tons		Quota %	
	1975-76	76-77	1975-76	76-77	1975-76	76-77	1975-76	76-77	1975-76	76-77
Centrale	5.460	5.480	600	750	3.280	4.110	300	400	9.1	9.7
Diallassagou	12.555	12.605	700	750	9.790	9.453	1.400	1.250	14.3	13.2
Koulogan	8.300	8.330	500	750	4.180	6.247	600	850	14.3	13.6
Baye	6.144	6.160	600	750	3.690	4.620	450	550	12.2	11.9
Kani-Bonzan	4.360	4.360	500	340	2.180	1.482	150	150	6.9	10.1
Ségué	2.600	2.800	450	730	1.170	2.044	160	200	13.7	9.8
Sokoura	6.675	6.680	500	400	3.340	2.672	240	200	7.2	7.5
Ouenkoro	2.500	2.600	500	730	1.250	1.898	200	400	16	21.1
	48.594	49.015			28.880	32.526	3.500	4.000	12.1	12.3

probable that they will produce a limited yield or go dry after a few years. Water is therefore the central concern of the population and a reliable water supply its most urgent need.

The sandy soils of the plain, while fertile when manured or fallowed, are subject to rapid nutrient loss from overcultivation. High-producing hybrid strains of millet and sorghum increase this danger. Interpretations of remote sensing data have indicated that continued cultivation creates expanses of bare soil which fallowing cannot regenerate (4). Wind erosion then continues desertification. The sandy soil of the plain is, however, ideal for millet cultivation, and its growth potential is more seriously limited by an uncertain water supply than by nutrient loss.

The cliffs are a series of nearly vertical scarps divided by gorges and valleys running the length of the plateau edge. Remnants of valley and cliff erosion form a talus slope which rings the cliff rim. Below the talus is a seasonally-inundated valley which collects runoff from above. The bodies of water thus formed dry up in December to February. Millet is cultivated on the cliff slopes on carefully-constructed terraces designed to hold the runoff water for as long as possible. Soil for these terraces is brought up from the plain below and carefully manured. Along the northeastern and eastern border of the plateau, water runoff would allow the installation of small reservoirs for supplementary irrigation. The soils here, however, are extremely poor and millet cultivation is in competition with pastoralism.

The Dogon

The Dogon (pop. 250,000-300,000) live on the Bandiagara plateau and the Gondo plain, some 300 kilometers southwest of Timbuktu. In the Sahelian zone, this region suffers a highly variable rainfall (28% annual coefficient) of between 300 and 600 mm. The uncertainty created by the oscillation of years of drought and years of rain and by uneven rainfall distribution within a given year makes planning for the future a condition of survival. It is understandable that the reduction of uncertainty has become a dominant theme in Dogon art and philosophy, the purpose of Dogon rituals, and a focus for their ingenious technology. This technology of risk reduction has two main features. Planting strategies are carefully adjusted to the contingencies of rainfall within each year; long-term strategy centers on storage.

Dogon planting strategies are a continuous decision-making process which responds to changing conditions during the rainy season. If early rains are available, the farmer will begin planting, but will hold back enough seed to sow again if the first rains are not quickly followed by regular precipitation. Depending on the success of his millet planting, the farmer will interplant beans (Vigna unguilata Volp), sorrel (Hibiscus abelnoschus esculentus), or the small and drought-resistant grain, fonio (Digitaria exilis).

Recently, cash crops such as peanuts have become a part of this risk-reduction strategy. Risks are no longer limited to productivity and consumption; they now concern the protection of economic investments in fertilizer, ploughs, and modern technology. One of the drawbacks of this trend has been reduced production of urgently-needed cereals.

Operation Mils

The entire area, as part of the Fifth Region of Mali, is administered from the regional capital of Mopti, a city known as the Venice of Africa. The Fifth Region is the center of activity of Operation Mils (hereafter Op Mils), one of several semi-autonomous agencies in the Ministry of Rural Development (5).

Op Mils was created with funds realized from the sale of emergency grain supplies sent by the United States during and following the 1972-1974 drought. It has since received substantial funding from USAID, but will have to show a profit in order to remain a viable instrument for change. One of the main goals of Op Mils is the increase of millet (Pennisetum typhoides) production. Millet is still the main staple of the rural as well as the urban population of Mali. Considerable increases in production will be necessary in order to create marketable surpluses for the cities and for deficiency areas such as the Sixth Region, to keep up with population increases, and to establish reserves needed for drought emergencies. As will be seen, major increases in productivity will be difficult to achieve over a wide area. Realizing this, Op Mils has extended its activities into more and more marginal zones in order to fulfill its commercialization quotas. It will be argued in this study that greater efforts toward reducing storage losses combined with a policy designed to increase productivity in a restricted, favorable zone may offer a more viable strategy.

To aid farmers in obtaining higher yields, Op Mils's extension agents have introduced improved seedbed preparation, planting in rows, weeding, seed treatment, and the use of ammonium phosphate fertilizer and urea. They are also involved in the construction of small dams and feeder roads, the promotion of onion cultivation on the plateau, and, more recently, a functional literacy program financed by USAID. Profits from the commercialization of grain are expected to finance these activities after the withdrawal of USAID support.

Op Mils and the Prevision System

Villages in Mali are assigned a prevision (quota) by the government indicating the amount of grain they must sell to Op Mils. The quota system is under the jurisdiction of the Ministry of Interior and Defense, which is also in charge of the distribution of food to deficiency areas and to urban centers. In the Fifth Region, the quota is fixed by the military governor and, at the local level, by the commandants du cercle. The system is therefore beyond the authority of the Ministry of Agriculture, OPAM, and Op Mils. The commandants du cercle, however, base their requirements on yield and production estimates provided by Op Mils. Villagers may appeal if they feel their allocation too high to fill or too low to permit them to sell enough grain to meet their taxes.

The prevision is pegged to the tax structure, which brings the Ministry of Finance into the process. This appears to be a basic reason why commercialization is extended into marginal areas; it is in the government's interest to have the largest possible number of farmers selling grain to pay their taxes. This creates pressure on small farmers who have not enough grain to sell, and it raises serious problems for development:

1) Increasing production in marginal areas is certain to produce serious erosion problems in the future;
2) Transportaion costs are too high and transport along sometimes impassable roads is close to impossible;
3) Commercialized grain not brought out of the area is lacking adequate storage facilities, subject to high losses due to insects and destruction during the rainy season;
4) Op Mils does not have the resources in personnel or funds to service the areas adequately; it needs its resources in key production areas (Bankass, Koro).

Individuals aged 16 to 65 pay a head tax of 1960 MF, the equivalent of 60 kg of millet at official producer prices of 32 MF/kg (6). Many farmers have little or no alternative source of income and are thus forced to sell their grain. Individual villagers, however, are not forced to sell to Op Mils; who will and who will not provide grain to the purchasing agents is a decision left to village councils and village chiefs.

Operation of the Prevision System on the Village Level

In January, villages begin to receive both empty sacks and money corresponding to their quotas. The sacks are distributed, generally one to every three tax-paying members in a family. This will net 1065 MF per taxpayer ($\frac{100 \text{ kg} \times 32 \text{ MF}}{3}$), about half the necessary head tax.

If the village has too little grain to meet the quota, attempts are made to purchase the missing amount on the markets of other towns or from individual farmers outside the quota system. Numerous villages claimed that they could supply about three quarters of the necessary grain, but that they could not meet the rest of the obligation. At Madougou Bore, for example, the quota was four tons, and the villagers said that they could procure three tons; at Petaka the quota was twelve tons, and the grain available, according to the villagers, was about nine tons. In some instances villagers said that they had sent a delegation to the *commandant* pleading for a reduction but had been refused. Now, however, they found the original quota "perfect. Not too large and not too small. Just perfect." It is difficult to judge the extent to which the implementation of the quota system exceeds the boundaries of a commercializing operation and becomes oppressive. It is reported from Dourou that "farmers say that their village chief is threatened with imprisonment if the quota is not filled" (7). The OPAM agent at Bandiagara justifies the size of the *prevision*: "Dogon farmers must be forced to sell because otherwise they will all switch to onion production and the country will starve" (8). Such statements support our own data and provide insight into the threats made and their rationalization.

Table 3a shows regional differences in land area available to individual cultivators; land area is one of the major differences between the high-production zones of Bankass and Koro and the marginal areas of Douentza. Although the percentage of active cultivators varies little from region to region, their productivity varies considerably (see Table 3b). As a consequence the grain available for auto-consumption per cultivator in Madougou (Koro) is more than double that per cultivator at Douentza. More seriously, the grain available after commercialization for auto-consumption per person is only half that assumed to be the national average (75.4 kg/year) (see Table 3c). Yet the amount of grain commercialized in marginal villages such as Petaka exceeds that of high producing villages such as Madougou.

Table 3a. Commercialization Quota Related to Surface Cultivated Yield and Production (Figures from Operation Mils)

Village Circle	Surface by Hectare	Surface Hectare by Cultivator	Yield - Kg Hectare	Total Production Kg	Commercialization Quota in Kg
Petaka (Douentza)	324.11	0.452	318.18	103.126	12.000
Madougou (Koro)	982.80	0.816	450	442.260	40.000
Dimbal (Bankass)	603	0.635	350	211.050	15.000

Table 3b. Cultivators as % of Total Population (Enquête Agricole, 1976/77)

Village Circle	Total Population	Working Cultivator			% of Cultivator of Total Population
		Men	Women	Total	
Pétaka (Douentza)	1208	390	321	711	59
Madougou (Koro)	1982	613	592	1205	61
Dimbal (Bankass)	1602	511	439	950	59

Table 3c. Millet Available to Farmer for Consumption and Storage After Commercialization

Village Circle	Production by Cultivator in Kg	% of Grain Commercialized by Op Mils	Available for Auto Consumption by Cultivator in Kg	Available for Auto Consumption (and Storage) per Person in Kg
Pétaka (Douentza)	145	11.64	128.12	75.435
Madougou (Koro)	367.02	9.04	333.82	202.96
Dimbal (Bankass)	222.16	7.10	206.37	122.38

To pay their taxes, villagers are obliged to sell to the government at its price (32MF/kg) in January and to buy back millet for home consumption in August (during the soudure) at the then higher official price (56 MF/kg). In order to do so, they are often forced to sell their livestock or to go into debt, thus compounding their marginal situation. The next year's crop is already partially claimed for payment of debt (9).

Millet-rich farmers, on the other hand, are unable to sell to the government at the January price, because the available quota must be divided by the village council among as many taxpayers as possible. OPAM funds are frequently insufficient to purchase both quota grain and the surplus of larger farmers. The system thus creates a disincentive to produce for large landowners while, at the same time, it causes shortages for farmers who cannot produce enough to feed their families (Table 3 a,b,c). To counteract this difficulty, a policy permitting smaller farmers to build up reserves and larger farmers to sell surpluses needs to be developed.

But the objectives of greater equity in income distribution and rapid increases in production can lead to conflicting policies, as can the goals of providing for the urban population and allowing farmers to build up their grain reserves. Yet, a sound development policy cannot depend solely on technological improvements - it must be based on social justice or it will fail in the end. This is well recognized by GOM officials as well as by AID officers who struggle with the problem.

U.S. support for Op Mils thus presents a moral problem difficult to solve at the local level. Should we refuse Mali funds needed to raise the standard of living of its rural population and to provide food as a security against drought because we see inequities in the Malian system? Can we attach conditions to our aid without seeming to interfere in the internal affairs of the country, without being told that we have similar inequities at home? USAID officials in Mali and elsewhere need guidance from Washington; how far should they go with a policy based upon human rights? Human rights should also be defined more broadly than the political suppression of individuals.

Yet it can be well argued that aid designed to increase production and productivity is the best way to help Malians rid themselves of oppressive grain- and tax-collecting practices. Increased production and productivity would allow greater food reserves at the village level and larger supplies to the marketing boards. Such reserves would engender a sense of security and a reduction of fear; farmers would cease to hide their grain from the government and the government would cease to seek it by force. But larger reserves will depend in part on better storage and the reduction of losses to pests. If current losses could be eliminated, enough grain might be saved to satisfy both farmer and government.

The Storage System at Collection Centers

After the quotas have been assigned and the millet amassed, farmers bring their grain (already shelled by the women) to Op Mils collection centers. The grain they bring is frequently the oldest available, or of a quality they do not wish to store themselves; farmers are reluctant to sell current harvests. Having been prepaid (in January with the arrival of the empty sacks) for their millet, and receiving no transport payment (which is rebated to the federations, local cooperatives, and not to the growers), farmers have little interest in rushing their grain to the collection centers. The delay between shelling and delivery considerably increases the likelihood of infestation in grain purchased by Op Mils. This likelihood is further augmented by the unwillingness of Op Mils to accept millet treated with ash in the traditional manner of Dogon farmers. Millet is therefore open to attack by *Trogoderma* and other pests even before it arrives at collection centers.

At the collection centers, storage space is insufficient and ill-designed. Quantities of several hundred tons of millet may be stored on open ground without protective cover. Grain may be kept in old storerooms and school buildings or in the residence of the extension agent, together with his goats and chickens. Even specially-designed warehouse space is often poorly constructed.

Common pest problems include:

1) *Trogoderma* (walls are covered with *Trogoderma*, and the ground may be covered with a heavy carpet of meal and *Trogoderma* shells);
2) *Ryzopertha dominica f.* (Capucin or Bostryche);
3) *Tribolium confusum d.*;
4) *Sitophilus granarius*;
5) *Sitotroga cerealla*;
6) Bats (construction in newer structures with corrugated iron roofs attracts bats, who, while they do not eat millet, infest the grain with their feces);
7) Rodents (rats, mice and squirrels -- a serious problem in some warehouses and in open storage);
8) Plant parasites (10) (the low moisture content -- 5-7% --

renders these a less-significant problem); and
9) Impurities (a serious problem in degrading the commercial value of the grain; a sample from a poor store at Diallassagou showed impurities of 12.5%).

Inexperienced storeroom management is often responsible for some of the considerable losses that occur. In one storeroom several tons of millet had been deposited in bulk and, after six weeks, showed so heavy an infestation of Trogoderma that it appeared to be crawling. Yet new grain was simply poured atop the infested stock.

Countermeasures are limited. Fumigation of the grain deposits is not economically possible because of the difficulties of access to remote areas. Extension agents use DDT or HCH, but these are often thrown onto uncleaned floors. Operatives cannot, however, be blamed for the generally unmanageable conditions. To improve the situation, Op Mils has begun construction of 28 warehouses, each of 50 tons capacity. The buildings, however, were originally designed for the storage of farm equipment and are ill-suited to their new purpose. One was so poorly constructed that it collapsed before completion (11).

We now turn to traditional storage practices among the Dogon. If the perpetual complaint of experts (expatriates and Malians) about OPAM and Op Mils storage is its lack of ifrastructure, this is not the case for traditional storage. This may be a prime reason why losses in farmer-owned granaries are much below those in Op Mils warehouses. We begin with a discussion of the cultural significance of the granary in the life of the Dogon. We hope to show that Op Mils can reduce its own losses by placing greater reliance on the traditional system, shifting some of its storage needs back to the private sector.

Traditional Agriculture: Storage Myth and Method

History

It is a harsh environment. Pushed by other, more powerful tribes, the Dogon gradually conquered their current homeland between the 13th and 16th centuries, defeating the earlier inhabitants, the so-called Tellem. The Tellem grew fonio (Digitaria exilis, a small-grained grass that can be cultivated without the use of iron implements) and stored this grain in small, delicately-ornamented clay granaries built into the dry, almost inaccessible caves of their sandstone cliffs, but the Dogon invaders brought with them

Table 4. Analysis of Stored Millet Samples, December 1976 M. Italo Dante de Murtas

Samples	Animal Parasites	Vegetable Parasites	Humidity	Impurity
Federation storage in Diallassagou (millet of 1976-77 after 2 months of storage)	Capucin or Bostryche: (very frequent) Rhizopertha dominica F. Coleoptera-Bostrychidae	_Aspergillus_ _sp_.	5%	3.5%
OPAM warehouse in Diallassagou (millet of 1974-75 damaged by rain)		_Aspergillus_ _sp_. _Fusarium_ _sp_.	7%	12.56%
Op Mils storage in Toroli (millet of 1976-77 commercialized)		_Aspergillus_ _sp_. _Penicillium_ _sp_.	5%	1.1%

a different, more productive grain, the millet <u>Pennisetum typhoides</u>.

Dogon myths illustrate the historical relationship between millet and fonio and between the Tellem and the Dogon themselves. Fonio, according to these myths, was the smallest and heaviest element in the universe. At the moment of creation the God Amma caused this stored power to explode and to form the prototypes of all existing things. The genesis of millet was different. The Dogon culture hero, the Nommo water spirit, descended to earth with the first grains, standing on a celestial granary and carrying with him the tools of the smithy and the knowledge of ironworking, the manufacture of the hoe.

The Dogon thus received a complete agricultural system as well as a model for accepting innovation and change. In the historical memory of the Dogan, the millet-iron- agro complex constituted an agricultural revolution. Their willingness then to accept outside aid is of great importance to the outcome of current attempts to introduce new technological innovations and production methods (<u>12</u>).

To cultivate millet, the Dogon needed more soil on their rocky cliffs, so they carried earth up from the plain (and called themselves "the tired people" because they placed earth where the God Amma had forgotten to put it). On the descending cliffs they constructed terraces, sometimes too tiny to support more than one or two plants; one can almost speak of millet gardening rather than farming.

Dogon methods of fertilizing their soil and cultivating their fields were sophisticated and effective. Households prepared fertilizer each year by mixing ash, dung, millet chaff, millet stalks, and refuse in a <u>fumiere</u> set to one side in the courtyard of the compound. The annual fertilization of the terraced lands on the cliff produced an intensive agriculture.

Millet agriculture produced enough wealth to permit the Dogon an architecture famous for its adaptation to the environment, its functional economy of form, and its simplicity and beauty. The most striking example of this architecture is the granary. Dogon granaries, built of rock, wood, and banco (clay with an admixture of millet chaff), are about 2 1/2 meters high on the cliffs and plateau and, on the plain 3 1/2 meters. These dimensions do not include the conical grass roofs that protect the granaries against sun and rain.

Figure 1 (left). Granary (Toroli, Koro) is being loaded with millet. A man inside the building is arranging the heads.
Figure 2 (right). Farmer showing millet.

Granary Construction

Granaries are sometimes constructed in long rows outside villages to reduce fire risks. They are built by heads of households and by traditional professional builders. (Opportunity costs and expenses for a granary range from 3000 to 8000 MF.) Nine rocks or logs form the foundation for a grid of strong branches on which the mud floor of the granary is laid, 30 cm above the ground. The granary itself is thus protected from the torrential waters that rush through a village during the rainy season. Its floors are strong enough to support 2 1/2 tons of millet, stored unthreshed. This may distribute weight, impede insect movement, and help aerate the grain.

Granaries on the plateau usually have two tiers and are partitioned into four sections. In these granaries millet is kept after shelling and is separated from other products -- beans, peppers, onions, and nonedible valuables, all arranged in a mythically sanctioned order (13).

The larger granaries on the plain are a more recent adaptation to an environment permitting greater production and more storage space. The Dogon could not settle on the plain until after the arrival of the French ended their wars with the Peul. Plain granaries can therefore be considered a relatively recent, innovative solution by an African population to new problems posed by a different environment. Because it is on the plain in the cercles of Koro and Bankass that production potential is greatest, we shall be concerned with this storage system.

The larger granaries on the plain have no partitions. Millet here is stored on the head according to a fixed schema: four rows in the rear, pointing away from the wall, and the remainder parallel to it at a distance of perhaps 1 to 2 meters. One row of heads is placed against the door, forming a kind of shield. The heads in this front row are sometimes damaged by termites (and even by birds) if there is no door. The termites, whose damage is usually limited to the outer layer of millet heads, are removed by scraping. More serious termite infestation occurs in the fields.

Granary openings were once protected by elaborately-carved wooden doors, many of which are now in the hands of European and American collectors. Carvings on these doors represented the ascending and descending generations of the Dogon family, ancestors, the water spirit Nommo, and turtles, crocodiles, and zigzag patterns recalling the cycles of rain and drought and of human existence. Today simple wooden pa-

Figure 3. Millet drying on rooftop.

Figure 4. Millet sorting and loading.

nels are used, although some have carved locks. I have noted that those granaries which have carved doorlocks and those which have religious symbols in relief on the banco walls are better maintained and show less losses.

Ownership

The ownership of granaries reflects the structure of Dogon society. The largest and most important are owned by the Ginna, the "great family" or lineage. It is stocked with grain of jointly-owned fields. From it comes the family's grain tax and aid to family members in need; it is a repository of wealth. It provides security against times of drought and reduces risk for individual members of the family. Heads of individual households also own private granaries as well as private fields. Household granaries tend to be smaller, and they serve the family's immediate needs during the year. Finally there are women's granaries, used to keep small quantities of shelled millet that will be eaten within a relatively short period of time. These granaries also store other household items.

Management

How effective are these granaries in fulfilling their functions? Some farmers claim that their granaries hold grain for generations without losses; others complain that their losses run to 60% annually. The anecdotal evidence is confusing. A clearer picture emerges from a study of Dogon grain storage as a total system.

Traditional farm storage demands not only structures such as granaries but also, and most important, careful management. The difference in the quality of care given privately-owned granaries and government-managed warehouses may well be an important factor in explaining the difference in losses in either system.

In the traditional system, quality control of stored grain is fundamental. Millet heads are sorted and classified during the harvest: very poor and aborted grain is not cut; poor grain is cut but kept apart; and good grain is prepared for storage. The millet is transported in 20-kg bundles by the men and in baskets of somewhat lesser weight by the women. The good grain is spread on the flat terraces of Dogon houses for drying, and the poor grain is held for immediate consumption. Farmers assert that their criteria for quality judgements include examination for insects that might contaminate the granaries.

Figure 5.

Woman pounding millet in wooden mortar. Next to her is the famous Dogon basket. One basket contains approximately 200 heads of millet and enough for a family of 10 people for a day. The basket is square at the bottom and round at the top signifying the world and the sky. Between them, contained in the millet is the very force of life and creation.

The very best quality millet is reserved as seed grain and is usually kept under the kitchen roof, protected by smoke against insect attacks. The remaining high-quality grain is moved into granaries six weeks to three months after harvest, in December or January. By then the grain is usually very dry, its moisture content no more than 10% (see Table 6).

Traditional Agriculture: Crop Losses

Beyond two preliminary surveys I made in the summer and fall of 1976, no previous study on crop losses in traditional granaries has been conducted in Mali (14). Storage of millet on the head seems largely confined to the Sahel, and there is no reference literature on the subject. Some research has been conducted in Senegal, and more extensive investigations are planned by the Tropical Products Institute (TPI) and the League for International Food Education (LIFE). Only a long-term study can give us the assurance and accuracy we need in order to be confident of the results.

This report presents the analysis of one hundred and forty heads of millet from fourteen granaries throughout the Dogon area. The analysis indicates an average of 13.65% loss per granary. This average is useful in estimating the amount of edible grain in relation to the total granary stock of a village, and it corresponds to our earlier findings (15). It is, however, the composite of a very few large losses and many rather insignificant losses. The average includes preharvest losses as well as postharvest losses in stored grain and does not represent an annual loss inside the granaries (see Tables 5 and 7). This high variability encourages the hope that peak losses can be reduced through the diffusion of existing techniques and the support of farmers who make use of traditional knowledge.

In 1976/77 our research concentrated on the millet-rich areas of the cercles of Bankass and Koro, around Douentza, northwest of the plateau, and around Korientze, near the inland delta of the Niger. Villages were chosen in regard to regional differences in grain production, in the availability of clay, wood and stone for construction, and in traditional grain-handling methods.

Within each village, granaries were selected from lists compiled by the Service Statistique. Information on these granaries could thus be related to the Service Statistique's complementary household and landholding data. Our main concern was to obtain a representative rather than a statistical sample. A farmer's granary is his checking account; permission to examine it is not readily given, and we were grateful

Table 5. Evidence of Losses Thought to be Prior to Harvest on Millet Heads Taken from Traditional Granaries

Regions	No. of Millet Heads	Losses in %					Year of Harvest
		Birds	Borers	Aborted Grain	Diseases	Total	
Douentza	10	0.75	0.20	4.20	0	5.15	1975/76
	30	0.33	0.89	9.25	0.90	11.37	1976/77
Mopti	10	0.80	2.05	6.20	0	9.05	1975/76
	30	1.20	1.07	11.42	0.07	13.76	1976/77
Bankass	50	1.25	0.30	3.27	0.15	4.97	1975/76
	20	0.32	1.32	17.65	2.97	22.27	1976/77
Koro	30	0.93	0.20	5.63	0.08	6.86	1975/76
	20	0.82	0.37	2.25	0.90	4.34	1976/77

Table 5a. Evidence of Losses Thought to be Post-Harvest on Millet Heads Taken from Traditional Granaries

Regions	No. of Heads Analysed	Losses in %					Year of Harvest
		Mice, Rodents	Termites	Sitophilus G. Sitotroga C.	Rot	Total	
Douentza	10	0.3	0.10	3.009	0	3.40	1975/76
	30	0.05	0.20	1.04	0	1.29	1976/77
Mopti	10	2.20	0	0	0	2.20	1975/76
	30	-	-	-	-	-	1976/77
Bankass	20	4.21	0.10	5.48	0	9.79	1975/76
	50	1.55	0.50	3.94	0	5.99	1976/77
Koro	20	0.92	0.05	6.05	0	7.02	1973/74
	30	3.25	0	4.40	0.73	7.78	1974/75
	30	1	0.03	3.09	1.56	5.68	1975/76
	20	-	-	-	-	-	1976/77

Figure 6.
Millet is kept in the fields before storage. Losses during this period may be considerable, although thorns keep out larger animals. The ground is covered with ash, which is thought to prevent some animals from attacking grain.

Figure 7.
Millet is pounded outside village in order to prevent chaff from entering houses. The large mounds of chaff are used in producing compost.

for the confidence that farmers offered us.

Selecting representative samples of millet heads from a full granary was difficult; the heads were tightly packed inside, and access to the center of each granary was almost impossible. For this reason, our sample tends to overrepresent certain easy-to-reach sections of the granaries. We assumed, however, that damage caused by some insects (e.g. Sitotroga cerealla) would be heaviest around the periphery of the grain stack, along the walls, and under the roof. Rodent damage is probably similar. Mice may find it more difficult, or unnecessary, to penetrate into the interior. As a consequence, samples from the easily-accessible outer layers of the stored grain may give a distorted picture of total losses. We selected ten heads from each granary, paying as much attention as possible to these problems. The selected heads were measured and examined for damage. The various pests leave characteristic traces, and their identification was therefore possible.

Identification of particular pests permitted us to distinguish preharvest from postharvest losses, but it remained difficult to isolate damage that had occurred after the actual harvesting of the millet but before transport to the villages. Some fields are up to 30 km from the villages of their owners. If there is much grain, if farmers are occupied with other tasks, or if farmers do not want to alert the government to the size of their crop, grain may be left harvested but in the fields for up to three months. During this period termites, rodents, and birds cause extensive damage. Rodent damage caused in the field is difficult to distinguish from similar damage caused inside the granary. Bird damage that occurs during the rooftop drying of the grain is difficult to isolate from the damage of earlier bird assaults. Appropriate and specific preventive measures will require, however, that such distinctions be attempted.

Preharvest Losses in Traditional Granaries

Heavy preharvest losses of millet have been reported for the Op Mils area. These losses cannot be estimated from an analysis of stored grain alone, because farmers rarely harvest badly damaged heads, and, when they do, they tend to reserve them for immediate consumption rather than storage. The relatively low preharvest losses evident in Table 5 can be attibuted to these factors and should not be read as representative of total in-field losses.

According to a study carried out for the Operation de Protection des Semences et Recoltes du Mali by Italo Dante de Murtas, Masalia lepidoptera noctuidas is currently the cause of widespread millet damage in the Mopti region. Infestation of 80-100% and harvest losses of 20% are reported. Other borers have also been discovered: Sesamia sp., Eldana sacharina, Chilo pysocaustalis, and Atheribona sorghi.

A variety of field fungi, such as Schlerospora graminola, Tolysporum penicillariae, and Spacelotheca sorghi, have been isolated, but fungi do not, for the moment, seem a major problem.

Table 5 shows a dramatic rise in aborted grain from 1975/76 to 1976/77. These figures can be explained by the sudden lack of rain in late August, 1976, when the millet was in bloom, and possibly by late and heavy rains during the 1976 harvest.

In other parts of Mali, heavy field losses due to rats are asserted by the Murtas study, but we have no specific information on the area under discussion. The lack of water may render such attacks less severe in the Seno-Bandiagara area. Rodent destruction of individual millet heads is easily distinguished; extensive sections are attacked and the glumes characteristically damaged. Heads attacked in the field may be gnawed on all sides, but those damaged inside granaries, where they cannot be easily moved, show attack on one side only.

Birds are responsible for extensive damage to millet, and in particular to early-maturing varieties planted near water (as in Fatoma, near the inland delta). The bird species most frequently blamed is known locally as queleaquelea; it attacks maturing grain. Major damage is also caused by wild pigeons, which attack the grain while it is drying on rooftops or lying in the fields after harvest.

Termites are a serious pest, both in the field and inside granaries. In the field they may envelop the millet head with their mud constructions. Inside granaries their attacks are localized near walls or doors. Extensive damage also follows recurring swarms of locusts of various species. Such damage is easily recognized by the clean-cut edge of the chewed-off kernel.

Modern methods of control have met with some success, but suffer a number of drawbacks. Current attempts to control Masalia involve double applications of Endosulfan in a method developed in Senegal. Despite, however, the extensive losses caused by Masalia, such treatments appear uneconomical at present; even if they are successful, the low price of grain and the low yields achieved by traditional means of production make Endosulfan too costly.

Seed-grain treatment with a Heptachlore-Thirame compound is more common. The treatment, however, has only recently been introduced by Op Mils, and there is considerable resistance on the part of farmers who feel it ineffective and too expensive. In our interviews farmers were suspicious of the overall benefits of herbicides and insecticides and question their effects on the environment (16).

Although little is known about the impact of toxic compounds on the fragile environment of the Sahel, current knowledge does warrant concern about the spread of pesticides, especially large-scale applications, as are occasionally effected in Mali designed to control birds or rats (17).

Traditional methods for controlling pests are not always practiced. Farmers are reluctant, for example, to burn millet stalks after the harvest, because millet stalks are vital for granary roofs, cooking fuel, construction, and basket-weaving. The disturbance to the ecological equilibrium caused by the Sahelina drought may underlie the dramatic increase of Masalia (which appears to be continuing) and the major explosion of the rat and bird populations. With the return of more favorable rainfall these pests may recede, thus reducing the need for major chemical intervention in the environment.

Postharvest Losses in Traditional Granaries

Rodent Damage. The severity of rodent invasions varies within the region from village to village and from year to year. Eruptions of rats or mice can be large enough to destroy the reserves of entire villages. In Bankass in 1976/77 the commandant du cercle complained that infestation in some villages was so heavy the "mice whistled in the trees like birds," yet in the adjacent cercle of Koro there appeared to be no rodent problem.

Our loss counts of rodent damage in granaries range from 0.3 to 4.1% and do not take into account such peak infestations. The loss figures are those in granaries under normal conditions, where rodent populations have leveled off. The picture is thus distorted for the region as a whole.

The Dogon combat rats with traps and even guns. Children also hunt and eat them. Villagers claim that this solves the rat problem but that children do no eat mice, "so we can do nothing about them." Well-kept granaries, however, are con-

Figure 8.
Granary window shows damage to heads. This damage is not representative of all heads inside.

Figure 9.
Floor of granary in January. Frass is indication of losses. The floor is not cleaned before reloading.

stantly checked for openings in the banco and for crevices around the doors. These are patched as a daily routine.

An OPAM expert indicated that rodent-extermination teams will be sent to heavily-infested areas. The effectiveness of such action remains to be seen.

Fungi Damage. Grain on the head is stored for up to four years, but in bulk with the addition of ash it is reported to keep for five to seven years. Five-year-old millet, however, even if undamaged by insects, is thought to lose its taste. This may be caused by *Aspergillus* and other fungi that will develop after five years of storage, even though the moisture content has remained low. Once developed, perhaps during a rainy season, the infection will continue to spread (18).

Insect damage. The most peristent and widespread loss of millet in traditional granaries is due to insect infestation, in particular to *Sitotroga cerealla* and to *Sitophilus granarius*.

The insect infestation we discovered in millet stored in traditional granaries may, however, have begun in the field either before or after harvest. Weevils (*Sitophilus granarius*) deposit eggs, which remain invisible, inside kernels. The Angouin moth (*Sitotroga cerealla*) deposits its eggs outside the kernel, but its newly-hatched larvae penetrate within. Kernels may appear sound and undamaged even while germ and endosperm are being consumed (19). Because the entrance holes of both *Sitophilus* and *Sitotroga* are invisible to the naked eye and only the exit holes of the larvae are obvious, farmers are unable to identify infested heads unless they actually see the weevils.

In our December/January research we found no living insects in the granaries, possibly because of the low moisture content (5-7%) of the grain at that time (see Table 6). We therefore counted the kernels with open holes on a 5 cm section cut from the center of each millet head. The rest of the kernels were treated for evidence of eggs or living larvae, but almost none were found.

Table 7 shows the percentage of holed grains on 10 heads taken from 14 different granaries and computes the average losses for all heads. The results show a remarkably skewed distribution that is summarized in Graph 1. It appears that the distribution of damage within each granary follows a characteristic curve no matter what the degree of infestation. Thus in the granaries examined, one to two heads contained

Table 6. Moisture Analysis - July/August - Nov/Dec 1976

Sample	Moisture %	Protein %
R202	9.05	10.7
R203	10.05	9.6
R204	10.09	10.2
R205	9.08	10.5
R206	9.09	10.4
R207	9.01	10.0
DTK3	5.00	(1976/77 grain)
OP07	7.00	(old grain destroyed by rain)
DI08	5.00	(1976 grain)

Operation Protection de Semences, et
Conservations des Recoltes - Bamako
Mr. Haco Dante de Murtas

R202-R207 were analyzed by Prof. Cho C. Tsen, Department of Grain Science and Industry, Kansas State University, Kansas. These samples were taken from granaries in Fatoma, Bankass, Doro, and Sanga in July/August. Prof. Tsen noted that samples showed no signs of reduction in protein content due to moisture retention.

Samples DTK3-DI08 were taken from storage rooms where grain was in bulk and sacks.

The samples are not strictly comparable and any hypothesis built on these measurements must be verified by further study.

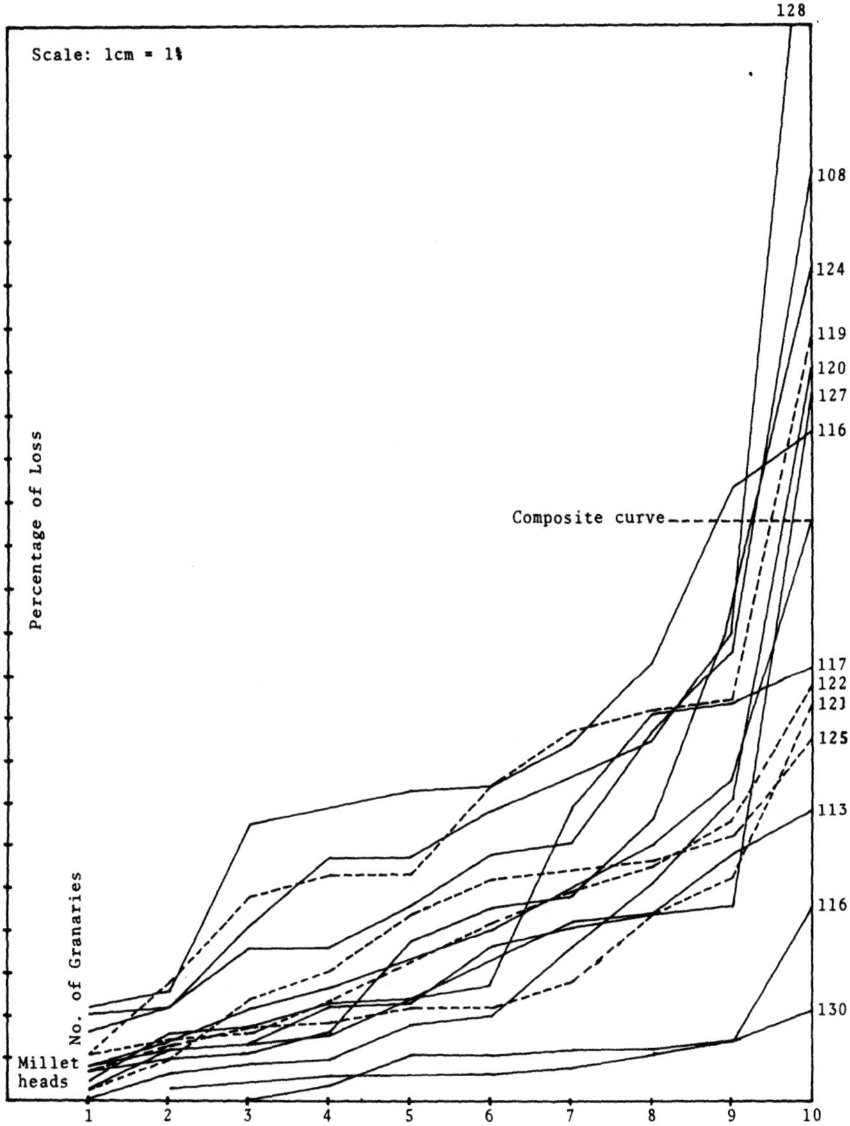

Graph 1.
Distribution curves of losses in samples of ten millet heads caused by Sitophilis and Sitotroga in fourteen granaries.

Table 7. Percentage of Grains Damaged by Sitophilis and Sitotroga on Ten Millet Heads from each of Fourteen Granaries

Granary No.	No. of Millet Heads										Average Loss
	a	b	c	d	e	f	g	h	i	j	
116	2.17	2.59	6.48	6.90	7.27	7.36	8.37	10.28	14.32	15.65	8.1
120	0.13	0.71	0.92	0.96	1.81	2.01	3.57	5.08	7.09	17.33	3.9
113	0.50	1.58	1.74	2.27	2.40	3.33	4.23	4.44	5.83	6.83	3.3
121	0.69	1.26	1.66	1.78	2.20	2.24	2.80	4.48	5.26	9.35	3.1
127	0.79	1.24	1.33	2.20	2.29	3.60	4.07	4.37	4.64	16.89	4.1
108	1.62	2.15	3.62	3.65	4.59	5.83	6.07	8.70	10.65	21.50	6.8
117	0.32	1.21	1.34	1.52	2.44	2.72	6.90	9.11	9.34	10.25	4.5
124	0.71	0.97	1.10	1.56	3.69	4.49	4.84	6.58	11.48	19.56	5.4
120	1.98	2.20	4.08	5.71	5.72	6.84	7.64	8.50	11.00	30.13	8.4
130	0	0	0	0.40	1.08	1.10	1.15	1.19	1.35	2.07	0.8
118	1.08	2.76	4.79	5.30	5.34	7.41	8.74	9.33	9.35	17.85	7.1
122	1.14	1.44	1.59	2.44	3.33	4.23	4.88	5.48	6.56	9.85	4.0
125	0.32	0.95	2.36	3.12	4.42	5.15	5.35	5.55	6.23	8.52	4.1
104	-	0.32	0.50	0.58	0.60	0.61	0.79	1.13	1.35	4.56	1.0

Percentages represent grains damaged on 5 cm sections cut from the center of each head.

Table 8. Coefficient of Loss in Four Granaries 1973/77 at Madougou (268 Millet Heads)

Year	No of heads	Weight of heads in grams	Average weight of each head in grams	Wt of grain off the head	Wt of grain per head	% yield of grain per head	Coefficient
1973/74	241	15,000	62.24	9,400	39.00	63	9
1974/75	234	18,000	76.92	12,100	51.71	67	5
1975/76	230	18,500	84.78	13,700	50.87	70	2
1976/77	264	19,500	73.86	14,000	53.03	72	0

Table 9. Grains Damaged by <u>Sitophilis</u> and <u>Sitotroga</u> on Millet Heads by Region and by Year

Region	Loss in % due to s/s	Total % Losses	% of losses due to s/s	Year
Douentza	3.009	3.40	88.50	1973/74
Koro	6.05	7.02	86.18	1973/74
Bankass	5.48	9.79	55.97	1974/75
Koro	4.40	7.78	56.55	1974/75
Douentza	1.04	1.29	80.62	1975/76
Bankass	3.94	5.99	65.78	1975/76
Koro	3.09	5.68	54.40	1975/76
Mopti	0.0	2.20	0.0	1975/76

Damage caused by <u>Sitophilis</u> and <u>Sitotroga</u> ranges from 56% to 88.50% of all damage to grain stored on the head inside granaries examined. The total damage, however, amounts to only 6-9% of grain stored as early as 1973.

about 50 pércent of all boreholes. This may be accounted for in different ways. One hypothesis relates infestation to the location of the heads within the granary; the heavily-infested heads would come from areas close to the roof, the doors, or the walls. This would be a reasonable assumption if most holes were caused by the moths. A second hypothesis relates the pattern to the life cycle of the insects and to their cyclical appearance during periods of higher humidity. Older heads could conceivably have suffered several periods of attack, while more recent heads have been exposed during only one season. The problem appears to us intriguing enough to warrant further investigation and a search for preventive measures.

Various experiments have been carried out to determine the correlation between insect damage and weight loss. In Uganda, experimenting with maize in a bamboo crib, J.C. Davies calculated that in a sample showing 10% insect-bored grain, the total weight loss was 2.7%. Davies also indicated that the percentage of bored grain in crib-stored, untreated maize in Uganda varies from 47-75%. Samples taken from bag stacks in store at a Ugandan plantation showed three-month-old maize to be 16% bored and crawling with insects. These figures represent a weight loss of 3.5-10%. In India, Venkat Rao et al. (1958) found the percentage of weight loss in sorghum grain to be a third to a half the percentage of grains bored by insects (20).

The Harris findings (1976) give a coefficient of correlation between progeny of weevils and weight loss of 0.96. Harris states that in an undisturbed lot of maize the number of weevils can be used as a reliable indicator of losses.

Because of the low moisture content of the grain during the period of our research, we did not encounter any weevil activity, and we could not duplicate the Harris experiments. We were, however, able to measure the progressive increase in the weight losses from 1973/74 to 1976/77 in a set of granaries owned by a wealthy farmer in Madougou (Koro). Results provided a coefficient of 0,2,5,9 between the weight of the grain and the weight of the heads of each succeeding year (see Table 8). Using the Harris coefficient, one could estimate the corresponding increase in weevils over the same period. It would, however, be preferable to conduct additional research during the rainy period, measuring the incidence of weevils at that time.

Table 9 summarizes the percentage of loss on grain heads by region and by year, and shows the relationship of borehole damage to total losses. The table allows us to establish a range of total losses between 3.4% and 9.8%. It makes no clear distinction between years or regions. We suggest a reason.

Although some Dogon farmers maintain their granary stocks according to years, they tend to mix the old grain with that of the new harvest. Granary 116, for example, contained some heads with 14-15% boreholes as well as some with 2-6%. This suggests a mixed stock of old and new heads, a possibility acknowledged by the farmer. Granaries 108, 107, and 118, with examples of 18-30% bored grain, present similar cases. Other interpretations are, however, possible. As the moisture level of the grain declines, insects cease to be active. As suggested above, controlled study over several years could show whether reinfestation occurs as old grain comes into contact with freshly-harvested, infected heads, or whether rising humidity in the granaries during the rainy season induces annually-recurring damage by insects.

The damage caused by Sitophilus and Sitotroga goes beyond their direct effects in traditional granaries, because these insects open the way to secondary pests such as Trogoderma and Tribolium confusum. These are present in large numbers in grain stored in government warehouses, but neither appears to be a problem in traditional granaries, in part because the unshelled millet inhibits their movement. We observed a case where 30-40% of the grain delivered to a Federation warehouse in Diallassagou had been destroyed six weeks after its arrival. Additional grain, purchased by the Parent-Teachers' Association for the school canteen, was thereafter mixed with the infested grain, creating a race between the Trogoderma and the children to see which could eat faster.

Although losses to pests in traditional granaries are less severe than generally assumed, they pose a problem for the marketing boards as well as for the individual farmer.

The question is not whether local storage is better or worse than marketing-board warehouses, but whether losses can be reduced in both to improve the system as a whole.

We shall now turn to some of the traditional techniques used by Dogon farmers to combat losses to pests.

Traditional Methods of Fighting Losses to Pests

Dogon farmers spend a good deal of their time taking care of their granaries. During the rainy season they protect their windward sides with mats, preventing the entrance of water and moisture. After the rains, a fresh coat of banco is applied outside walls, repairing whatever damage has been caused by heavy rains.

Farmers realize the advantage of relining the inside walls as well, but this is rarely done, because it involves extensive work in emptying the granary and it reduces interior space. Although granaries are close to empty after the soudure, the newly-harvested grain is placed atop whatever is left inside.

The most widespread technique for the protection of grain within the granaries is smoking. Millet chaff mixed with pepper is burned, and its smoke flushes out rodents as well as insects. The granaries are relatively airtight, and this method of fumigation appears effective.

The leaves of plants such as Andropogoneos and Combretum (vines), whose odor is thought to repel both rodents and insects, are placed with the grain. More important, however, is the use of the ash of Boscia Senegalensis and millet stalks. It is scattered on the floor and rubbed into the walls of the granary. Ash is also commonly used as an admixture with threshed grain. Millet thus mixed may be stored in baskets and jars or simply on the floor and shelves of the plateau granaries.

Such traditional techniques as smoking could become more effective through appropriate technological innovations. We suggest for example a small pottery stove with an iron grid that could be placed inside a granary. New and improved methods should be developed and tested in cooperation with the Dogon farmers.

In addition to traditional use of ash and smoke, inert dusts (not currently used in Mali) may offer a relatively inexpensive solution. Such dusts are not favored in the United States because they reduce the commercial grade of the grain and its test weight, but these considerations should be less important in Mali. Inert dusts include diatomaceous earths, silica aerogels, and activated clays.

Research on the use of inert dusts has been conducted in India and Kenya. It is in the semi-arid tropics, where humidity does not pose a major problem, that these inert dusts can be applied to local level storage problems.

Farmers have recently begun to purchase HCH, which they throw into granaries to reduce insect damage. HCH is sold by Op Mils for mixture with seed grain and is probably ineffective when applied in this way, as well as too costly. The practice may also produce harmful side effects, which need to be investigated (17).

Recommendations

Further research on losses in traditional granaries and on traditional methods of grain protection during storage is needed. A project collating available information on these issues and evaluating their reliability is now being conducted by Dr. Peter Golob at the Tropical Products Institute. In terms of the granaries in the Dogon region and elsewhere in Mali and the Sahel, it will be particularly important to follow the development of infestation on particular millet heads over an extensive time period--perhaps three years. The source (and sources) of infestation needs to be pinpointed (field, previously infested grain, granary walls, other) in order to adopt better preventive measures. Some emphasis should be placed on the following questions:

- signs of infestation in field
- signs of infestation when grain is placed into granaries (sometimes the delay can be two-three months)
- period of heightened insect activity (temperature and humidity changes)

The study should also look at the management of traditional granaries over a number of years. Observations should determine:

1. When and how often a granary is opened and by whom. Method of cleaning after grain has been removed. Method of loading grain and opportunities for checking this point.
2. The rate at which losses occur. The criteria of loss by which farmers decide to sell grain or continue to store it.

Further research should be linked to a program that would try to reduce farmers' losses by diffusing traditional Dogon techniques throughout the area. Traditional prophylactic techniques such as the use of ash and smoke, or building maintenance such as the renewal of roofs, the relining of inside walls, and the sealing-off of holes and spaces around doors are high in opportunity costs and can be expensive in terms of local incomes. (A new roof would cost about 2000 - 3000 MF.) If farmers could store grain belonging to Op Mils, they could be given credit for such repairs. Maintenance costs would be negligible and would probably reduce storage costs for Op Mils.

In order to encourage farmers to make use of these techniques, a training program for granary hygiene should be instituted. The director of Op Mils has discussed with me the possibility of training his agents in traditional methods as a beginning step.

Future Possibilities for Grain Storage

Farmer Storage

Family-stock storage and seed-grain storage is an important element in the production and consumption cycle. Our research showed that farmers feared to sell grain until they had a base line security stock for their own needs, preferably of more than one year and up to three years. In 1976 we found no stocks older than three years, although some villages claim that prior to the drought they held grain for up to seven years. After five years it was felt that the flavor of the grain began to change, and it became unpalatable. I would estimate from conversations with farmers, however, that losses to insect pests may well rise dramatically after three years. Farm storage must therefore be improved if the farmer is to see any advantage in holding his grain for more than three years. At present he can probably maximize his gain in selling after two years (see Table 10).

Recommendation. We recommend a policy that allows farmers to build up their stocks for a two-to three-year period:

a) it would reduce the risk to the area in times of drought and reduce the need for increased national reserves;
b) it would give time for a more systematic and efficient organization of international aid in the case of drought than was evident during the last emergency; and
c) it would encourage farmers to sell their surplus if prices are favorable and thus be an incentive to production.

The lack of local grain reserves, on the other hand, is a disincentive to production. Farmers are forced to sell at unfavorable prices, and they cannot establish their own security.

Op Mils Storage

CRED addresses the problem of Op Mils storage as one of official marketing rather than from the perspective of the farmer. CRED calculates that two-thirds of official marketing volume requires storage at the major centers and that storage at intermediate points should provide only for accumulation of adequate quantities for transshipment.

Given current circumstances (outlined below), this is a reasonable policy for Op Mils. Because grain prices are fixed in advance and because Op Mils does not sell to the population at collection points, storage there means higher

Table 10. Evolution of Value of One Cultivator's One Year Millet Production Stored in a Granary from 1973-74

Year	Initial Production	Percent of coefficient of loss	Weight of Loss in Kg	Actual Weight of stock	Price per kg	Actual value of stock in MF
73-74	333.82	-	-	333.82	20	6,675
74-75	333.82	2	6.63	327.14	32	10,468
75-76	333.82	5	16.69	317.13	32	10,148
76-77	333.82	9	30.04	303.78	32	9,721

expenses and losses.

Op Mils is currently engaged in increasing the productivity of the farmer; increasing total production; and commercializing grain at the farm level and transmitting it to OPAM. With these goals, Op Mils finds no benefit in storing grain at the local level; the longer it holds grain, the greater the cost of storage and the larger the storage losses absorbed by Op Mils. It is, rather, in the interest of Op Mils to transfer its grain to OPAM as rapidly as possible and as close to the point of collection as possible (because OPAM transport rates -- 2.863 MF/kg -- are fixed without regard to distance). Op Mils is not currently selling grain at the local level and does not profit from storage by holding grain for OPAM.

Op Mils officials, therefore, do not seek long-term storage facilities; but demand a system that can move grain rapidly into OPAM hands. The current infrastructure of roads and transport is insufficient and presents Op Mils with its most serious obstacle to effective operation. Lacking a suitable infrastructure, Op Mils needs, for the immediate future, short-term storage facilities to counter the high risk that it will be forced to hold considerable grain during the rainy season. For this we recommend open-air storage on plinths.

Nationally, the relationship between Op Mils and OPAM should be reconsidered and duplication of effort eliminated. Op Mils should handle distribution of grain to the population and develop a storage system at the arrondissement and cercle levels that would permit a profit with which to finance development. Op Mils should then be encouraged, through adjusted transport rates, to carry surpluses (amounts in excess of emergency grain reserves and annual needs) directly to Mopti to be stored. If these conditions could be brought about, Op Mils could develop more permanent storage and, at the same time, begin to lease traditional granaries for its own use. It would also be profitable to construct its own storage facilities at Mopti to handle bulk grain.

OPAM Storage

CRED believes that storage facilities available to OPAM are roughly adequate in capacity if not, perhaps, in quality. They suggest an annual rate of increase of 3-5% (which corresponds to their hypothesis of 2.5-3% population growth rate with some development of urban and export markets). This is based on the assumption that Mali has reached close to self-sufficiency in cereals and increases in production are needed only to balance population increases (21).

Storage, however, cannot grow at the rate of 3% annually; silos do not grow this way, nor are aid projects funded in this fashion. It is therefore understandable that Malian officials are pressing for the immediate construction of larger facilities to assure future capacity. In discussions with FAO and GTZ experts, OPAM has voiced the following objectives in requesting 28,800 tons of storage capacity:

1) to meet the most urgent storage needs for the annual marketing of grains (short-term storage of one year and, in some cases, medium-term storage of eighteen months) -- 74% of the total storage to be constructed (21,300 tons);

2) to cover long-term storage needs for the FAO/AO security stock project -- 26% (7500 tons) (22).

This storage would be constructed with German Assistance and would consist of warehouse-type facilities. Its advantages include relatively safe storage in the major centers (where losses can be reduced by fumigation, (23) by efficient management, and by quality control of incoming shipments) and its use as an emergency reservoir in times of drought.

These are valid advantages. Is OPAM, however, the most efficient agent of change, or should Op Mils assume responsibility in the Fifth Region?

The transfer of grain from the commercializing agency (Op Mils) to *cercle*-level OPAM stores involves unnecessary additional handling. OPAM responsibility should begin only at the regional level, where grain can be held for long periods and where export decisions can be made (24). OPAM and the government would continue to determine the amount of grain to remain in *cercle* stores; transfers of grain ownership would merely be made on paper rather than in movement of the grain itself. In reducing the transfer of grain, losses could be cut. A standard of 2% loss is currently assumed on all received shipments; an additional 2-5% storage loss is later recorded. If the grain remained in the local area, these figures would shrink.

Reorganization of OPAM with French funding is now underway. German proposals for the construction of OPAM warehouses are also under consideration. Competition among international donors -- the Americans backing Op Mils and the Europeans, OPAM -- is not in Mali's interests and should not be allowed to develop. Donors differ not only in their approach to economic strategies, but also in their technical solutions. One such solution is examined below.

The Fatoma Granary: An "Invisible Technology" Solution

Following discussion with Ronald Levin (CDO, Bamako) in 1975, I undertook the design of a granary allowing:

1) construction by local masons and with local materials in the aesthetic of traditional architecture;
2) a 30-50 ton capacity and easy grain removal;
3) interior subdivision to permit the separation of different cereals and the separation of grain by owner;
4) an airtight structure requiring fumigation only when partially empty;
5) the minimal use of cement and other expensive materials;
6) replication by trained local labor; and
7) simple maintenance.

Such a building fitted into my concept of an "invisible technology" (developed in the construction of water-storage systems on the Bandiagara cliffs) (see Table 11).

At the suggestion of the governor of the Fifth Region and the director of Op Mils, the village of Fatoma was selected as the pilot site for such a granary (25). Fatoma (population 2000) lies on the route between Mopti and Douentza, accessible to government officials and other observers. It is the center of an *arrondissement* of 53 villages and is known as a village that "seeks progress". Its dominant ethnic group is Peul, and its inhabitants cultivate millet and rice and breed cattle.

After consultation with the villagers and the *chef d'arrondissement*, a granary site was chosen in Fatoma's marketplace, a location which would give the greatest possible prominence to the building. Construction began on September 3 and ended on October 15, 1975. Local laborers and masons built and helped to design the structure, a large, nearly cubic building. My hopes for an unremarkable structure were realized; visitors to Fatoma scarcely notice it among the traditional houses and mosques, although it occupies a place of honor. The pointed buttresses that sustain its walls and help support the weight of the millet inside provide welcome niches for local market women and give the granary the appearance of a traditional mosque.

Within are three compartments (two of 11.5 tons and one of 6.5 tons capacity), their walls lined with ferro-cement (chicken-wire and cement). A humidity-proof paint maintains low moisture levels during the rainy season. A large external staircase made of banco leads to the roof, composed of pre-stressed concrete tiles that sit on concrete beams. The roof is slightly tilted and is bordered with the traditional

Figure 10. Inauguration of Fatoma granary.

clay waterspouts that grace Mopti architecture; both insure adequate runoff of rain. On the roof are three raised chutes through which the three interior sections can be stocked. Each of these sections has a sharply-inclined floor which spills the grain into cement boxes in a small vestibule behind the front door. These boxes operate on the chicken-feeder principle; grain is simply scooped out of them and into bags.

The Fatoma granary has not yet been properly tested. It has never been filled to capacity; the asphyxiation of pests was therefore not expected. The first grain poured into the three compartments was contaminated, and this contamination spread until the building was fumigated. Thereafter, spoilage ceased. The Fatoma granary continues to be used by the Parent-Student Association and is monitored by the OPAM-TPI team of Barker and Gilman.

Free-Market Handling of Grain Surpluses

In the transformation of millet from subsistence to cash crop, grain storage has moved from a central position in traditional religious belief to an equally crucial place in the economic and political philosophy of the modern state. Grain marketing and storage policy are a matter of ideological commitment for Sahelian governments, their populations, and donor agencies; they define the limits of specific social goals and political and economic interests. The overall social philosophy of the Malian government favors the even progress of the peasantry over the creation of a priviliged elite and favors the development of competing government operations over a system of competitive free enterprise.

In his report to USAID, Helman (26) opined that: "Storage losses at the farm level are high, even in areas of dry climate, where storage techniques are well adapted to local conditions... Increasing permanent storage at or near the village level will only increase the inventory held in physical proximity to the farmer, at increased public cost, without really affecting his marketing options..." Our research results can be interpreted to reach different conclusions. We believe that, for the Fifth Region, village-level storage is least costly in terms of loss and that farmer inventories, which provide a sense of security against drought, must be increased. The first priority of the subsistence farmer is that he be able to sell without fear, not that he be able to sell.

Helman continues to identify transport from farm area to collection points, urban markets, and deficit areas as a cen-

Table 11. Cost, Management and Losses in Different Storage Systems

	Butler bins	Warehouses	Butyl silos	Open air storage	Traditional Granaries
Capital outlay US $ per ton (variables are size and transportation costs of overseas materials)	$30	$30-50	$25-75		$5-10
Level of administration of supervision required	very high	medium	high	low	low
Infestation losses	very low to 0	very high	low	fairly low	fairly low
Condensation losses if grain not very dry at time of loading	low, if drying system exists such as rotation	none	can occur if no precaution is taken	none	none, normally

tral problem; he concludes: "The principle objective of cereal marketing initiative is to attain increased storage of cereals in market proximity to the farmer, i.e. owned by the farmer, a village grouping, or agricultural operation subject to the farmer's influence, but in physical location as close to the central point of collection or market destination as practicable. This places the farmer in a better position to respond to market options" (27).

The Dogon farmer has at present few, if any, marketing options. Although the OPAM monopoly is restricted to the period of commercialization, free trade is discouraged, often with police action. In any case, the high cost of transport out of the area and overproduction in areas where farmers have adopted modern techniques under Op Mils guidance depress local prices below the official market price. Only a radical change in government policy will change this situation.

These are constraints to Helman's proposal and to the USAID project design paper's assumptions, not objections to their desirability. If the government were to restrict OPAM to a regulatory role (entering the market when prices are too low, withdrawing when they become too high), and if a system of village markets were established to stimulate production and trade throughout the region [markets can precede production increases and act as an incentive to farmers (28)] then farmer-owned village storage would be a viable option and indeed a desirable one. A suitable intermediate technology appropriate to such a strategy needs to be developed.

The ancient religious symbols carved into the doors of Dogon granaries and the sophisticated butyl rubber linings of modern granaries are both part of a struggle between man and pests, a struggle which must have begun with the first agricultural revolution, the first grain harvests. Losses to pests have serious consequences for humanity; they are both unacceptable and avoidable. In considering the protection of grain, we must also consider the protection of the small, traditional farmer and his cultural heritage. Only a humanistic and just policy will help man triumph over pests.

References and Notes

1. Some of the postharvest loss data presented in this paper were collected with the assistence of Hamidy Hama Diallo of the Operation de Protection des Semences et Recoltes au Mali, between December, 1976 and January, 1977 under USAID grant BKO 688-77-06. I have profited greatly from discussion at the AAAS Symposium on Postharvest Losses, Denver, and from comments on the paper as it was presented at ICRISAT, Hyderabad, at the Institute for Development Studies, University of Nairobi, and at IRAT, Bamako. All comments are gratefully acknowledged. Parts of this paper were submitted in a report to USAID, and I thank Ronald Levin (CDO, Bamako) for permission to publish them. I am also grateful for the help of Moriba Sissoko, Directeur du Cabinet in the Ministry of Agriculture, Bamako, M. Soumare, former director of Operation Mils, and M. Toure, current director of the Operation. The help of Mr. Quincy Bembow, Agricultural Officer, USAID, Bamako is acknowledged.
2. The CRED study, completed before the 1976 census, assumed a 6 million base population for 1976, a growth rate of 2.2% from 1976 to 1980, and a growth rate of 3.0% from 1980 to 2000. CRED then estimated that total consumption of coarse cereals would rise from the present 960 thousand tons to 1454 thousand tons by the year 2000. This would require a production of 1745 thousand tons. One CRED hypothesis assumes that by the year 2000 per capita consumption of coarse cereals will decrease in Mali by two-thirds (from 160 kg/pc to 53 kg/pc) as consumption of rice expands. To project a change from millet to rice consumption may be justified on a national basis; the preference for rice can be established in the larger towns and cities if the price difference between the two can be reduced. For the Dogon region, however, cultural preference for millet may be too strongly entrenched. Here millet is a way of life. "We cannot work on rice," farmers told me. "It does not stick to your stomach." Large increases in rice production may also depend upon Mali's ability to establish an export market for its rice surplus. This market is currently controlled by Southeast Asian rice, preferred for its texture and, in adjoining countries, less expensive than Malian rice.

   Figures published by CRED give a nationwide deficit of 22,500 tons of millet and 10,900 tons of rice for 1974-75, but predict a deficit of 100,000 tons of cereals for 1977-78 (60,000 tons of millet and 40,000 tons of rice). Listed in order of deficiency are the following zones: GAO, Mopti, Bamako-ville, Kayes. In contrast to CRED, the Institut d'Economie Rurale believes that the balance of millet as a factor in autoconsumption has gone from a surplus in 1971-72 of 32,700 tons to a deficit in 1977-78 of 21,500 tons (a differential deficit of 54,200 tons); with rice, this presents an overall deficit of 23,600 tons. Thus, they perceive a trend of increasing deficits of 24,000 tons annually. As causes, they list a series of developments: the relative decrease of agricultural labor in the total population (caused by the rural-urban migration); the rise in nonproductive consumers; an insufficient increase in the productivity of farmers (emphasis ours); an increase in cash-crop activity (too strong in relation to previous factor); too little progress in increasing yields of millet and sorghum (which receive only a weak secondary advantage from fields fertilized for cash-crops); and finally, a rise in auto-consumpton.

3. Table 1 shows evidence for sources of error in production estimates:
   1) Extension agents assert that reported increases in surface areas under millet cultivation do not reflect the traditional rotation pattern and that newly-cultivated fields are merely replacements for other areas left fallow.
   2) Individual extension workers, who see it as their job to produce increases in output and commercialization, experience a conflict of interest in making yield and production estimates. It is difficult to see how ecological differences within a small administrative district can account for the large increases and decreases in yield per hectare evidenced by the statistics outlined in Table 1.
4. R. Fanale, Utilization of ERTS-1 Imagery in the Analysis of Settlement and Land Use of the Dogon of Mali, M.A. Dissertation, (Catholic University, Washington, D.C.,1974).
5. The operational headquarters of Op Mils is now in Mopti. From that city, its activities extend across the cercles of Bankass, Koro, Bandiagara, and Douentza, and into parts of the cercles of Djenne and Mopti.
6. In 1977 the official government price to the producer is 32 MF/kg, a 60% increase over the 1974 price, but in view of inflation, near the same level of purchasing power. The price may be further increased for the campaign of 1977-78, according to a high official of the Ministere du Development Rural.

   L'Agriculture au Mali recommends that the price for millet be raised by 3 MF/kg, disregarding the question of a competitive advantage with prices linked to export goals because the authors are opposed to the export of cereals.

   If the prices of peanuts and cotton remain unchanged, profits from all agricultural products would exist in a better balance, but raising only the producer price of millet would be insufficient unless grain and agricultural policies were redefined.

7. Ritta Eskadrinen, personal communication.
8. Jerry Johnson, personal communication.
9. A case study (of the arrondissement of Bore) is available in Guggenheim and Fanale, 1976, Wunderman Foundation.
10. Three lots of millet were analysed for plant parasites. See Table 4.
11. One warehouse has been allocated to each zone d'extension rurale (ZER):

Mopti-Djenne	8 warehouses (not yet begun);
Douentza	5 warehouses (not yet begun);
Koro	7 warehouses (3 completed);
Bankass	8 warehouses (3 partially completed).

12. I have seen something mysterious, let us go there...
    I have seen something mysterious, I have come
    to tell you
    What is this?
    That is the new world that has come...
    -Dogon song-

Dogon culture is fascinated with the unknown, with change, with agriculture, and with technology. "We will try anything," a Dogon village chief told me. "Bring us a new seed or a new tool, we will try it."

The Dogon regard outside help with appreciative interest, but also with cautious calculation of risks; they weigh the advantages of new technology in relation to their own needs. The acceptance of new technology, however, confronts the Dogon farmer with cultural and social risks. The jealousy of others must be met should the farmer succeed. Fear of jealousy is so strong that women sometimes bring in their harvests at night to avoid the eyes of their neighbors. Acceptance of new technology must be confirmed by social leaders to avoid risking family relationships. The "great house" and its fields are held and controlled by an "elder brother." He must decide whether a new venture might impinge on the rights of the family; he bears ultimate responsibility for the welfare of all. When a USAID official and an Op Mils extension worker asked an old farmer, whose low-yield field was opposite that of a pilot farmer with an apparent bumper crop, whether he would join the Op Mils Program, they were surprised to hear him defer his decision: "I must ask my elder brother." The answer was no, because the risk of social discord was too great. The farmer added that the reason for the other's good crop and his own poor result was rain. Sharp difference in yield due to an eclectic rainfall pattern are possible in the Dogon area; that a single year's crop in his neighbor's field was a success was not proof to the old farmer of superiority of the new methods.

13. The ritual significance of these granaries has been discussed in a masterly paper by Jacques Flam, African Art Symposium, Columbia University, 1974.
14. My early research was largely confined to the questioning of farmers and the analysis of samples taken from their granaries. Interviews were conducted on the cliffs near Kundu in Sanga and on the plain in Baye, Bankass, and Koro. On the basis of this preliminary work we concluded that losses were generally in the neighborhood of 12-15%, but that they could reach 50-60% in some granaries (Guggenheim and Fanale, 1976, Wunderman Foundation).
15. This average, it is important to note, is not of an annual loss, but is a cumulative figure covering several years.
16. Cf. Kenneth Ruddle and Ray Chesterfield, "Change perceived as man-made hazard", Development and Change, 7, 3, 1976.
17. Pimentel and Goodman, in Survival in Toxic Environments, (Academic Press, New York, London, 1974).
18. Mary B. Hyde, in Storage and Cereal Grains and Their Products, M. Christenssen, ed. (1974).
19. R.T. Cotton and D.A. Wilbur, in Storage and Cereal Grains and their Products, M. Christenssen, ed. (1974).
20. D.W. Hall, Handling and Storage of Food Grains in Tropical and Subtropical Areas, (FAO, 1970).
21. The maintenance of the status quo of poverty and deprivation pegged to current productivity and consumption levels, added to a greater population density and an increasingly skewed distribution of wealth among richer and poorer farmers could lead to serious problems in the future. CRED's assumptions, which foresee no real growth but only the adjustment of supplies to population growth, add to the concerns voiced above.
22. Of this total, 4900 tons are allocated to the Fifth Region (Mopti 3500, Bankass 700, Koro 700). Agroprogress, Grain Storage Policy and Stock Rotation for Sahelian Countries for Phase III (Agroprogress GmbH, Bonn-Bad Godesberg, 1976).
23. Geoffrey Gillman and Darryl Barker of the Tropical Products Institute have done much to improve OPAM storage and to reduce loss. Bref Rapport sur l'état des entrepôts utilisés par l'OPAM à Sikasso, Segou, San, Mopti et Gao. G.A. Gilman, Bamako, July 1975.
24. French experts argue for export increases and a means of guaranteeing prices if surplus production is achieved. L'Agriculture au Mali, on the other hand, advises against the export of millet. I believe that the gradual build-up of a buffer stock, held in bulk in modern silos, could give the Malian government a year-to-year flexibility in deciding this issue.

25. The construction was financed by the Wunderman Foundation with the support of Christian Aid and other donors.
26. H. Helman, Cereal Marketing in Mali (1976).
27. The Helman Report has suggested that the Dogon farmer was eager to sell and that it was unprofitable for him to hold on to his grain. Dogon farmers throughout the area who were interviewed on the subject protested: "money is spurious and one cannot hold onto it; grain in the granary is worth more." Table 10 indicates this. We calculated the granary to hold about 2 tons, which meant a loss of 180 kg (5760 MF at current prices). In spite of the losses inside the granary, the farmer had not lost money by holding on to his grain, even though grain prices have not kept pace with inflation (20-23% per year). Since this is the case, he would be able to make a profit and could maximize his benefits by selling after one and a half years after which losses begin to mount more rapidly. This would give him enough money to invest in repairs to his granary and give him a greater margin of security.
28. Research on this topic is being undertaken by Mathias Von Oppen at ICRISAT.

8

Environmental Aspects of World Pest Control

David Pimentel

Introduction

Yields of valuable food crops are substantially diminished by a variety of plant and animal pest species that destroy crops during the growing season and also during the postharvest stage. Indeed on a worldwide basis, preharvest food losses to pests are high and are estimated at about 35% (1,2). Losses in the United States are estimated at 33% (2,3). These losses include destruction caused by insects, pathogens, weeds, mammals and birds. Although mammal and bird losses appear to be more severe in the tropics and subtropics than in the temperate regions, these losses are still low compared to the destruction caused by insects, pathogens and weeds. Added to this are the postharvest losses, estimated on a worldwide basis to range from 10% to 20%. In the United States, these losses are slightly lower and are estimated to average about 9% (3). The major pests of harvested foods are microorganisms, insects and rodents.

In an effort to reduce crop destruction by pests both bioenvironmental (nonchemical) and pesticidal control methods are employed. In the U.S., bioenvironmental controls for insects are employed significantly more often (about 9% of agricultural acreage) than are insecticidal controls (about 6% of agricultural acreage) (2). To control plant pathogens, some form of bioenvironmental control is used on about 90% of the crop acreage compared with the use of fungicide on only 1% of the acreage. For weed control bioenvironmental controls, primarily mechanical and cultural, are used on about 80% of the acreage, while about 17% are treated with herbicides.

The use of bioenvironmental controls appears to be

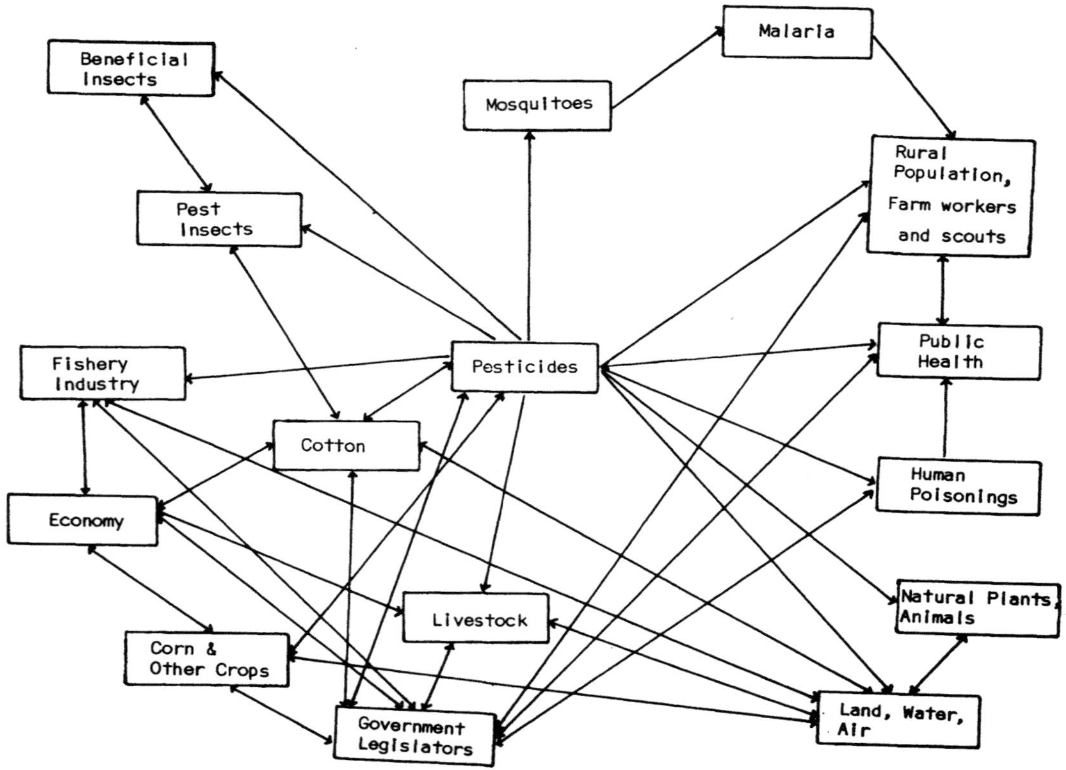

Figure 1. An "overview" of the major factors interrelated with the environmental problem of pesticide use in Central America.

more widespread in parts of the world other than the United States judging by the amount of pesticide used (4). Pesticides certainly play a significant role in world pest control but are again secondary to bioenvironmental controls.

In the world today nearly 2 million metric tons of pesticides are applied annually, or about one half kilogram per person (5). About 34% of this is applied in North America, about 45% in Western and Eastern Europe, and the remaining 21% throughout the rest of the world, primarily in developing countries (6). Estimates are that by the year 2000, the annual quantity of pesticides applied in the world will total more than 3 million metric tons.

Unfortunately pest control strategies, especially those relying heavily on pesticide, may have a detrimental impact on the environment. The three examples discussed below illustrate the range of problems that may be caused by pest control programs. The Central American and U.S. examples also are illustrative of the complexity of the environmental problems that pesticides may cause when they are used to control a pest species.

Environmental Problems Associated with Cotton Pest Control in Central America

Agriculture dominates the economy of Central America. Almost 60% of the population is involved in agriculture and an estimated 70% of the exports are food and fiber products. As for the value of export crops, cotton closely follows coffee and therefore is a dominant part of the Central American economy (Figure 1).

An estimated 393,000 ha are planted to cotton in the Central American nations of El Salvador, Guatemala, Honduras and Nicaragua. Clearly then, cotton not only dominates the economy but cotton acreage accounts for a major part of the farmland.

Cotton originated in Central America as well as in the West Indian Islands. In addition to being the native home of the cotton plant, Central America is also the native home of some of the most serious cotton pests including the boll weevil and several bollworm species (7). Because some of the natural resistance to insect pests has been lost in cultivated cotton, and temperature and moisture in Central America are optimum, conditions are ideal for the survival of the major cotton pests. As a result, serious pest outbreaks have occurred and often powerful insecticides have been used for control.

In nature several species of natural enemies (predators and parasites) function effectively to limit cotton-pest insect populations. However, when insecticides are applied without knowledge of both the ecology of the pest and beneficial insect populations, outbreaks of other pest species can occur. This has happened in Central America and is one of the reasons that more and more insecticide treatments have been required to deal with a worsening insect pest problem (7).

Insecticides Used in Cotton

The pesticide problem in Central America is primarily an insecticide problem; relatively small quantities of fungicides and herbicides are applied to cotton. In Guatemala where insecticide use is the heaviest, an estimated 80 kg is applied per hectare. Per capita use is one of the heaviest in the world.

Admittedly cotton would be difficult to produce in Central America without insecticides because of the severe insect problems there. Starting about 1950, when insecticides were first used extensively on cotton, about 8 sprays per season were effective. Since then the number of applications has risen to about 30 and in some areas reaches as many as 40 per season. The reason for the widespread increase is that insecticides represent the only currently available response the cotton farmer has to deal with both the insecticide resistance that has been developing in pest insects and the loss of natural controls (beneficial predators and parasites). As a result of the large number of heavy dose treatments, the cost of cotton production has risen significantly. However, insect pest control is still not adequate. Thus with production costs rising and yields reduced by pests, an economic crisis will continue to worsen unless sound pest management practices soon are implemented.

Human Poisonings

Reported human poisonings total about 3,000 annually in Central America with about 10% fatal (7). Methyl and ethyl parathion are the insecticides primarily responsible for most of the human poisonings. Most human contact with the insecticides occurs when farm workers and "scouts", workers who monitor pest and beneficial insect populations, come in contact with the insecticide-treated cotton plants. Often the insecticide residue is picked up while farm workers are weeding or harvesting cotton. The scouts are also exposed to toxic dosages of the insecticides while monitoring pest and beneficial insect populations.

Malaria and Mosquitoes

In Central America the anopheline mosquito (Anopheles albimanus) is the prime vector of malaria. The heavy use of insecticides on cotton growing in the coastal plain regions has also contaminated the aquatic habitat of the mosquito. The use of aircraft applications of cotton insecticides has compounded the problem since 50% to 60% of the insecticide applied by aircraft drifts outside the target cotton field (7). Thus most of the coastal plain habitat including the mosquito's aquatic habitat has been contaminated with cotton insecticides, and the mosquito population slowly has become increasingly tolerant to a wide variety of insecticides including propoxur. Currently, all insecticides, with the possible exception of landrin, are relatively ineffective against this mosquito.

This high level of resistance has resulted in a significant increase in mosquito control expenses. For example, house spraying with DDT costs only $1-$2 per house. After the mosquitoes developed resistance to DDT, propoxur, a carbamate insecticide, was substituted for DDT, which increased the cost to about $11 per house. With many mosquito populations now resistant to propoxur, the only potential insecticide available is landrin, which is projected to raise the costs to about $22 per house. Since houses have to be treated about 4 times a year, this has become a major expense for the people.

The more critical situation, however, is the fact that with ineffective control of mosquitoes, malaria, Plasmodium falciparum, is increasing. For instance in El Salvador, one of the worst problem areas, the incidence of malaria increased from 33,000 cases in 1973 to about 66,000 cases in 1974 (7) and unless some successful control measure develops, the number of cases will continue to rise.

Effect of Insecticides on Livestock, Other Crops and the Fishery Industry

Beef cattle that are produced on the coastal plains where cotton is produced may become contaminated with insecticides (7). Beef containing greater than 5 ppm of DDT, for instance, cannot be exported to the United States. In the past several shipments of beef have been "dumped" or could not be exported. Some of this contaminated beef is sold in the Central American market at lower prices (7). A few cattle have been killed following accidental exposure to high insecticide dosages.

Milk and other dairy products may contain high pesticide residues with concentrations averaging about 20 ppm and ranging as high as 100 ppm (7). Furthermore, beef and dairy cattle accumulate insecticide residues from both contaminated forage and contaminated concentrate feed, such as cotton seed meal. The forage is easily contaminated by atmospheric drift via aircraft applications.

When corn, one of the staples in the South American diet, is grown adjacent to cotton areas, the corn has been contaminated with insecticide residues. In addition, circumstantial evidence suggests that corn stunt disease is more prevalent on corn grown in the cotton regions, again because of the destruction of natural enemies that are important in controlling one of the insect vectors of the corn disease.

The yield of beans, squash and other food crops may be affected by insecticide pollution because the necessary pollinating insect populations have been reduced by the insecticide treatments. Crops grown in the coastal plain with cotton also contain insecticide residues. For example, from 0.1 to 10 ppm of DDT has been found in beans (7). The full extent of new insect problems that have resulted because of insecticide residues present in the environment has not been determined. Some preliminary reports suggest that in general insect pest problems are more severe on crops grown in the cotton region.

Reports have also been received that commercial fish and shrimp populations have been reduced due to severe insecticide contamination. Since accurate population data on fish and shrimp are not available, it is not possible to confirm or deny these claims. However, there is no question that insecticide residues do occur in both the shrimp and fish populations (7).

Fresh water and salt water (estuaries) are readily contaminated since pesticides drift in the atmosphere and alight on the water. In addition, insecticides are washed from the land into streams, ponds and estuaries.

Effects of Insecticides on the Environment

The water, land and air all become contaminated with insecticides when large quantities of insecticides are applied, especially when most is applied by aircraft. As mentioned, only 40% to 50% of the insecticide applied by aircraft lands in the target area; hence, most of the insecticide drifts off into the atmosphere and contaminates the environment (air, water, soil and biota).

The biotic environment in Central America probably consists of an estimated 100,000 species of plants and animals (7). Because of the intense biological activity of insecticides, it is anticipated that the large quantity of toxic insecticide applied to the Central American environment is having a significant impact on many animal and plant populations. Without good population data before insecticides contaminated the environment, it is impossible to give an estimate of the impact. On the basis of experience in the United States, the impact will be significant (See section on Environmental Impact of Pesticides in the United States).

Biological Control Backfires in the West Indies

In 1870 the Indian mongoose, (*Herpestes auropunctatus*), was introduced into Jamaica for rat control in sugar plantations and subsequently was introduced into Puerto Rico and other West Indian Islands (Figure 2; 8). Although initially the mongoose was highly effective for rat control, the benefits soon were offset by complaints from sugar plantation owners that rat control was ineffective. Complaints also came from small farmers who said the mongoose was preying heavily on chickens. Then in the early 1950s the mongoose was also incriminated as the vector and reservoir for rabies and human leptospirosis (9,10,11).

Why was the mongoose reported effective in rat control and later reported ineffective? From the evidence the mongoose was apparently highly effective in controlling the Norway or brown rat, (*Rattus norvegicus*), but ineffective against the tree or black rat, (*Rattus rattus*).

The black rat is believed to have been introduced first into Trinidad, British West Indies, about 1658 and was brought into Puerto Rico soon after that date (12). The Norway rat was introduced probably some years after 1700 (Figure 3). By 1877, when the mongoose was brought to Puerto Rico, there is little doubt that the rat population on the island consisted of both species of rats (Figure 3). The Norway rat was probably the dominant species, for as Zinsser (13) reports wherever the Norway rat has gone, "it has driven out the black rat and all rival rodents that compete with it."

The mongoose had ample opportunity to attack and control the Norway rat in the sugar plantations, since the Norway rat is a ground-nesting animal. Once the dominant

Figure 2. The Indian mongoose imported for rat control in sugar cane plantations.

species was controlled, the black or tree rats could easily become dominant (Figure 3). The mongoose, a nonclimber, however, could not reach this species and so it grew to be a pest. Support of this hypothesis comes from data of Pimentel (11) who reported trapping only black rats on farms where the mongoose occurred, but in towns and cities where the mongoose did not exist, the rat population was mostly Norway.

In addition to preying on chickens and being a vector of rabies in the West Indian Islands, the mongoose destroys ground-nesting species of birds and ground-inhabiting lizards. The reduction in lizard population is reported to have resulted in an increase in a sugarcane beetle (14). This rather simple example of biological control illustrates not only how far reaching the effects of a control method can be, but also how important it is to study the entire ecosystem before introducing a new species for any purpose.

Environmental Impact of Pesticides in the United States

An estimated 200,000 species of plants and animals exist in the United States (15). These species are vital to our survival, for we cannot survive with only our crop plants and livestock. No one knows how many species of plants and animals can be exterminated or how far the ecosystem can be altered before humans themselves are in serious jeopardy.

In 1975 more than .6 billion kg (1.4 billion pounds) of pesticides were produced (Figure 4) and about 360 million kg (800 million pounds) were probably applied to our environment to control about 2,000 pest species (16). If these poisons hit only the target species, there would be no pollution problem. Unfortunately less than 1% ever hits the target pests themselves (17). This problem is more complex when pesticides are applied by aircraft for often as little as 25% to 50% of the chemical ever reaches the target crop (18-22). The significance of aircraft applications contributing to pesticide pollution is clear when evidence shows that about 65% of all agricultural insecticides are applied by aircraft (23).

Alterations of the Ecosystem

The direct toxic effect of a pesticide on a pest or beneficial species is relatively easy to assess. Effects on the natural environment, however, may either be direct or be more subtle and hence go undetected for long periods

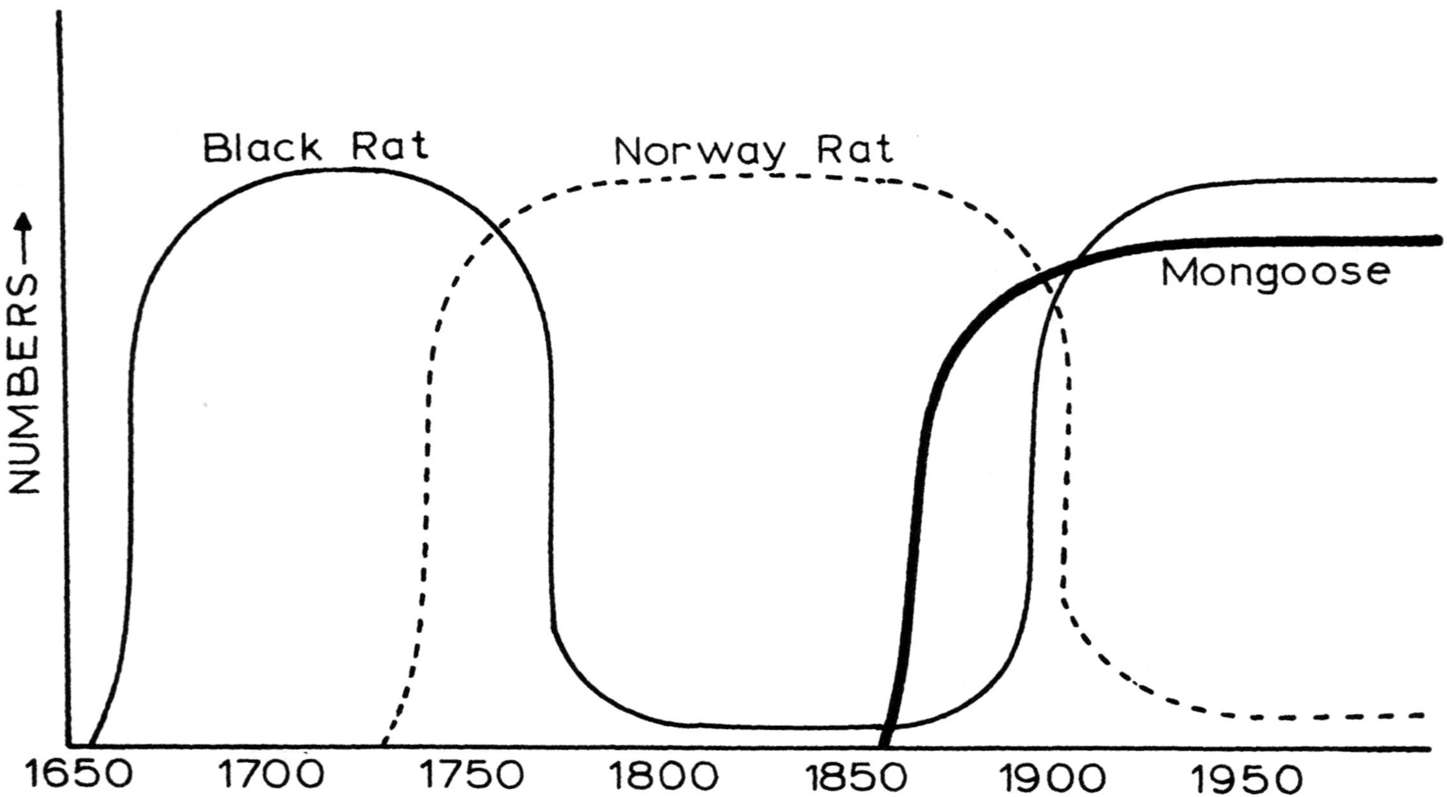

Figure 3. Graphic relationship of relative abundances of the mongoose, black rat and Norway rat in Puerto Rico from 1650-present.

of time.

Mention was made of our dependence upon the 200,000 species which make up the life-system in the United States. We depend upon this great variety of species for the maintenance of a quality atmosphere, for adequate food, for the biological degradation of wastes, and for other necessities for our survival. In the presence of sunlight, plants take in carbon dioxide and water and release oxygen, which is needed by humans and other animals. The oxygen (as both oxygen and ozone) also screens lethal, solar ultraviolet rays, keeping them from reaching the earth's surface. In addition, the plants are food for many animals, passing their life-making elements (carbon, oxygen, hydrogen, nitrogen, phosphorus, etc.) to the animals in the food chain. Eventually, microorganisms feeding on the dead and other wastes release vital elements for reuse by plants. In this way, all species of the life system interact and function to keep the life-making elements recycling in the environment.

With quantities of pesticides and other pollutants increasing, some species populations of the life system may serve as "indicator species". Their response will "tell us" when pollutant dosages are reaching dangerous levels in the environment.

Although pesticides are aimed at one or at most only a few species in biotic communities, usually these toxicants have some impact either directly or indirectly on most species present. Pesticides may influence the structure and/or functioning of entire biotic communities. For example, when the animal community associated with Brassica oleracea (cole plant) was exposed to either DDT (0.28 kg/ha) or a combination of parathion and endrin (0.22 and 0.20 kg/ha, respectively), significant changes in the community were observed (24). Initially, there was no significant difference in the number of herbivorous taxa present in the untreated and both treated plots (23, 20, and 19 taxa, respectively). Although the parasitic and predaceous taxa were quite similar in both the untreated and DDT communities (13 and 11, respectively), the parasitic and predaceous taxa in the parathion-endrin communities were practically extinct (only 2 remained). The loss of natural enemies in the parathion community was due not only to the direct destruction of many natural enemies but also to indirect destruction caused by a reduction of their food host and prey populations to such low levels that the parasite and predator populations were unable to live.

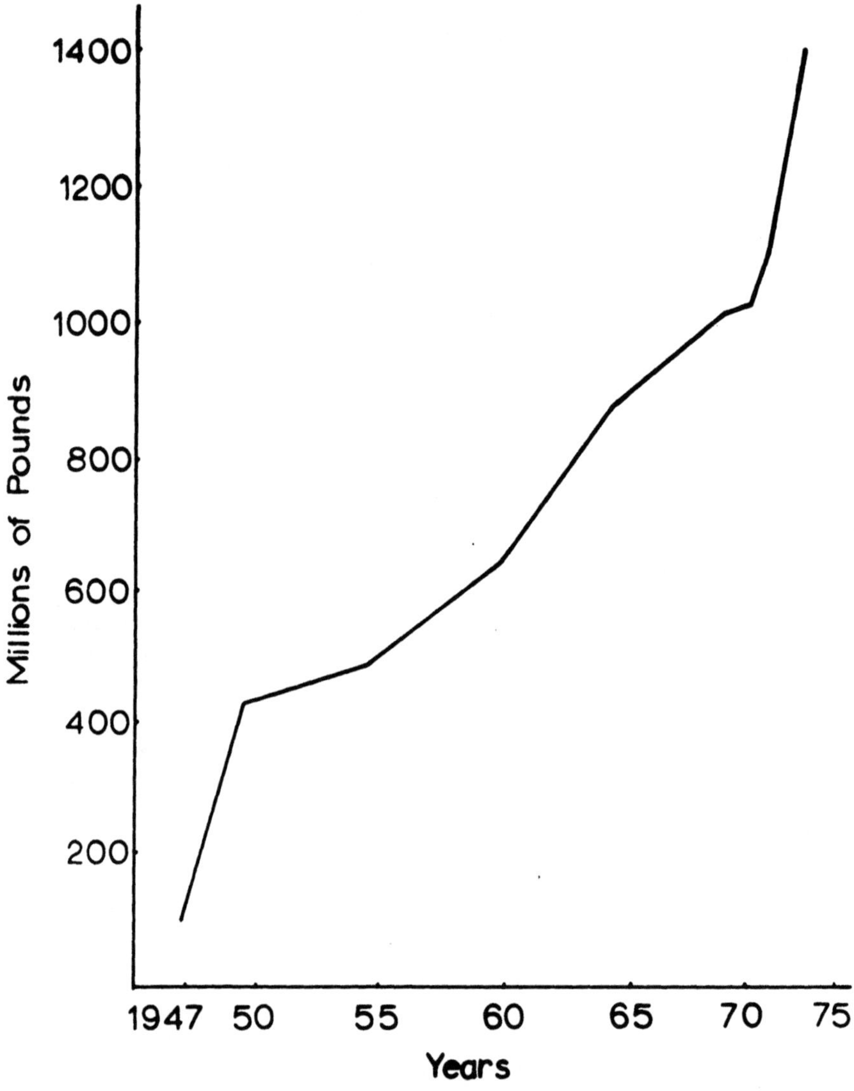

Figure 4.
Estimated amount of pesticide produced in the United States (16,54).

Densities of the various taxa per week in the three communities varied significantly (Table 1). Note that an aphid outbreak occurred in the DDT community and also that a large number of parasites attracted by the aphids were present. Meanwhile the parathion-treated community had been severely reduced in kinds of taxa.

Interestingly the ratios of parasites to hosts and predators to aphid prey were quite different in the three communities (Table 2). Aphid-parasite numbers were up but lepidopteran parasite numbers were down in the DDT community compared to the untreated community. Both, however, were higher than in the parathion community. Predator to aphid prey ratios were low in both the DDT and parathion communities.

Recognizing the complexity of interactions in the *B. oleracea* community (Figure 5) and the different degrees of

Table 1. The average taxa density per week per 125,000 sq. cm. recorded in three experimental communities (after 24).

TAXON	CONTROL	DDT	PARATHION
Aphids	162.7	1,106.3	20.3
Lepidoptera	12.0	4.4	1.0
Flea Beetles	1,107.1	3.5	1.5
Herbivores	643.5	3.4	1.0
Parasites	27.8	481.7	0.4
Predators	8.9	7.6	0.7

Table 2. The ratio of parasites to hosts and of predators to aphids in various experimental communities (after 24).

TAXON	CONTROL	DDT	PARATHION
		Parasites	
Aphids	9	34	1
Lepidoptera	57	22	8
		Predators	
Aphids	4.4	0.6	1.2

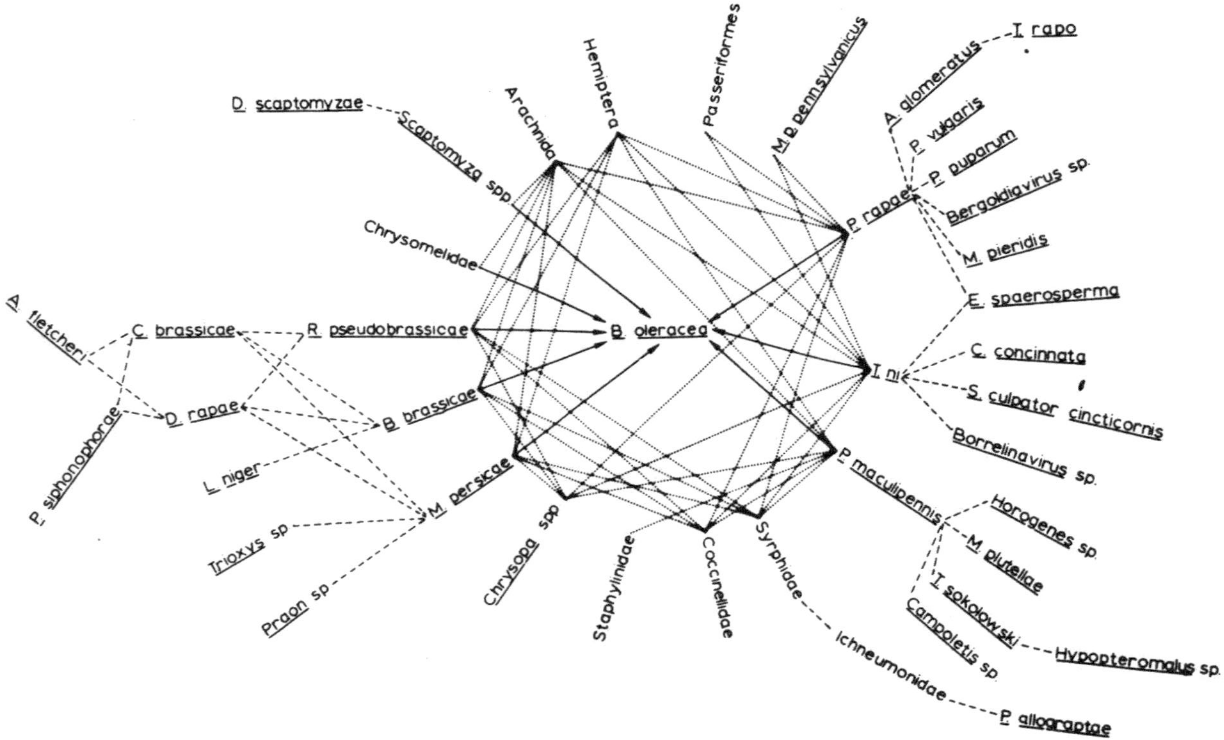

Figure 5. The relationships between the cole crop-plant (Brassica oleracea), the insect pests (——), the parasitic (----) and predaceous (....) enemies of the insect pests (25).

susceptibility that the various species have in the community (25), it is at times impossible to predict exactly how any one community will respond to the impact of a pesticide. Generally this community responded to pesticides by: (1) a reduction in total species; (2) population outbreaks of some species (aphids); (3) species low in the food chain increased once natural enemies were eliminated; and (4) predators and parasites high in the food chain were more severely affected than those species low in the food chain.

In another experiment a reduction in species diversity occurred in a fescue meadow when it was exposed to 3.4, 10.1, and 30.2 kg/ha of the herbicide, sodium cacodylate (26). Although recovery of species diversity of the plants was relatively rapid, it was not complete by the end of the growing season. Of greater significance was the fact that simplification of community structure resulted in substantial reductions in biomass throughout the growing season. For example, there was an 80-100% reduction in total biomass of all species except fescue within two weeks after herbicide application.

Reducing some species of stream insects on which fish feed may have a striking influence on the diets and perhaps the survival of fish. For example, crayfish were not in the diet of brook trout before treatment of a forest with 2.8 kg/ha of DDT. Immediately after treatment crayfish accounted for 99% of the trout diet (27). How this change in food affected the survival of the trout population was not investigated.

Herbicide destruction of plants upon which animals depend for food may cause significant reductions in animal numbers. Thus, 2,4-D applied to a gopher habitat reduced the green forbs by 83%, eventually resulting in an 87% reduction in the dependent gopher population (28).

These examples illustrate the influence of pesticide pollutants on ecosystems and the interdependence of all species in the earth's biological system. Plants, humans and other animals are all functioning parts of the same system or "establishment"—if one part of the life system is lost, the entire ecosystem, including man feels the impact.

Population

The application of pesticides directly to croplands, forests and other habitats may reduce and sometimes temporarily exterminate not only the pest, but also some nontarget species in the treated region. This is not surprising for pesticides are highly effective poisons applied specifically to destroy animal and plant pests. Although the direct effects of applications are relatively easily observed, the indirect effects are difficult to detect. For example, how do we discern whether numbers of a nontarget species are declining because of the indirect effects of pesticides or because of the environmental factors such as weather, which affect natural populations? In studying the indirect effects, the precise mode of dispersion through the environment and the dosage that reaches the nontarget species are difficult to determine.

Investigations as to why some raptorial bird species were declining in habitats where chlorinated insecticide residues were present illustrate the complexity of the problem. Wildlife biologists suspected that DDT and other insecticide residues were having an adverse effect, but recognized that urbanization was also contributing to bird mortality. Proof that DDT and other insecticides contributed to the observed decline of some raptorial birds required the exposure of predaceous bird species to known amounts of pesticides under controlled environmental conditions. To accomplish this, the American sparrow hawk was fed measured amounts of DDT and dieldrin or DDE to make body levels as high as those found in nature (29,30). Feeding the sparrow hawk these chemicals did cause the birds to produce eggs with significantly thinner shells and this contributed to a loss of eggs greater than that experienced by untreated controls. The laboratory evidence substantiated the field observations, leading to the conclusion that in some areas DDT and DDE were contributing to population declines.

The decline of lake trout in Lake George and other nearby lakes also illustrates the complexity of measuring the indirect effects of pesticides. For several years previous to and during the decline in the lake trout populations, about 4,535,900 kilograms of DDT had been applied yearly for pest control throughout the Lake George watershed. A portion of the DDT found its way into the lake, but the amount was probably small. Although DDT was found in both adult lake trout (8 to 835 ppm of DDT in fat) and their eggs (3 to 355 ppm), the mature lake trout appeared unaffected, and their eggs hatched normally. The reason for the decline in the lake trout population remained a mystery until it was discovered that the young fry were highly sensitive to certain levels of DDT in the eggs (31). Fry died at the time of final absorption of the yolk, just when the young were ready to feed. Mortality in the fry was significant at 3 ppm and, at levels of 5 ppm or more, mortality was 100%. The reason for the decline in the lake trout population in Lake George was then obvious.

On the California Channel Islands there appears to be an association between premature pupping of the California sea lion and organochlorine pollutant residues (32). Premature births have been occurring with increasing frequency, starting as early as January. The early premature pups are not furred and apparently died soon after birth. The later premature pups lacked motor coordination, breathing was short, and only parts of the lungs were inflated. DDT and PCB residues in blubber fat of females having premature pups averaged 824 ppm and 112 ppm, respectively, whereas residues in females producing full-term pups averaged only 103 ppm (DDT) and 17 ppm (PCB) (32).

Beneficial predaceous and parasitic insect populations have been reduced and even eliminated in some regions after insecticide has been used to control a pest species. When this occurs, then insect and mite species previously controlled by these beneficial populations can increase to outbreak levels. For example, when predaceous coccinellid beetles and other predator and parasite populations were unintentially eliminated in areas treated with DDT, chlordane, and other chemicals, outbreaks of mites (33,34), aphids (24), and scale insects (35,36) occurred. Sometimes the densities of these plant pests increased 20-fold above the control level usually achieved by their natural enemies.

Herbicide Increases Insect and Disease Problems

The herbicide, 2,4-D, has been applied to more acres for weed control in the United States than any other herbicide (37). Of all the crop acreage corn received the largest quantity of 2,4-D. In a recent study, it was documented that 2,4-D increased insect and pathogen pests on corn (38).

In 1974 field tests with normal use dosages (0.55 kg/ha) of 2,4-D, corn leaf aphid populations numbered 3116 per tassel compared with only 1420 per tassel in the untreated corn field (38). In addition, European corn borer attacks on 2,4-D-exposed plants were significantly greater (70%) than on untreated plants (63%).

In controlled laboratory studies with the corn borer, Oka and Pimentel (38) reported that corn borer pupae were 22% larger on 2,4-D-exposed plants than on the untreated corn plants. Moths from these larger "2,4-D pupae" produced 36% more eggs than moths from the smaller pupae from untreated plants.

In laboratory studies with two plant pathogens of corn, Oka and Pimentel (38, 39) reported corn exposed to 2,4-D had significantly more southern corn leaf blight lesions and had significantly larger corn smut galls than untreated corn. Again the 2,4-D dosages employed were normal use dosages.

The results of these investigations demonstrated that increased risks of attack by insects and disease to corn may result from herbicide treatments. Similar results have been demonstrated with other crops and other herbicides (39,40-51).

Interestingly, since 1945 increased losses due to attacks of insects and pathogens have been reported for crops in spite of greater efforts in pest control (3,52,53). How much, if any, of this increased loss caused by pests is due to the ecological impact of herbicides on crops is unknown, but evidence suggests that herbicides may be a contributing factor.

CONCLUSION

Considering all the environmental problems caused by pest control methods, pesticides appear to be the most serious hazard. Ample evidence suggests that current methods of pesticide use are an important hazard to biota in agricultural and natural ecosystems. To summarize, some pesticides have influenced the structure and function of ecosystems and communities and reduced species population numbers in certain regions. They have altered the habitat under some conditions; changed the normal behavioral patterns in animals; stimulated or suppressed growth in animals and plants; and increased or decreased the reproductive capacity of animals. Some pesticides have altered the nutritional content of foods; increased the susceptibility of certain plants and animals to disease and predators; and even changed the natural evolution of species populations in some regions. Although more information is needed on the impact these economic poisons are having on nontarget species, there is sufficient evidence for concern.

Based on available information, some generalizations can be made about the effects of insecticides, herbicides and fungicides on populations and communities of agricultural and natural ecosystems:

1. Pesticides tend to reduce significantly the numbers of individuals of some species in biotic communities. This has an ecological effect similar to that caused by a reduction in the number of species.

2. A significant reduction in the number of species in a community may lead to instability within that community and subsequently to population outbreaks because of alteration in the normal check-balance structure of the community.

3. After pesticide applications the species populations most likely to increase in numbers are those in the lower part of the food chain, that is, the plant feeders. This is, in part, because the parasitic and predaceous enemies which naturally help control numbers of plant feeders often are more susceptible to pesticide pollution effects.

4. In addition, any effective loss of species or intense fluctuations in number of species low in the food chain may adversely affect the dependent predator and parasitic species at the top of the food chain. This in turn further disrupts the structure and ultimately the stability of the natural community.

References

1. H. H. Cramer, Pflanzenschutznachrichten 20(1), 1 (1967).
2. D. Pimentel, Bull. Entomol. Soc. Am. 22, 20 (1976).
3. U.S. Department of Agriculture, Agricultural Research Service, Losses in Agriculture, Handbook No. 291 (Government Printing Office, Washington, D.C., 1965).
4. Production Yearbook 1972 (Food and Agriculture Organization, Rome, 1973).
5. W. Turtle, personal communication.
6. D. Pimentel, Energy Use in World Food Production, Environmental Biology, Report 74-1 (Cornell Univ. Ithaca, N.Y., 1974).
7. "An environmental and economic study of the consequences of pesticide use in Central American cotton production," Central American Research Institute for Industry, United Nations Environment Programme (ICAITI, Guatemala, 1977).
8. E. D. Colon, Datos Sobre la Historia de la Agricultura de Puerto Rico antes 1898 (privately printed, San Juan, Puerto Rico, 1930).
9. E. S. Tierkel, G. Arbona, A. Rivera, A. de Juan, Publ. Health Repts. 67,274 (1952).
10. R. H. Yeager, Personal communication.
11. D. Pimentel, J. Mammal. 36, 62 (1955).
12. J. G. Myers, Empire Marketing Board No. 42 (London, 1931).
13. H. Zinsser, Rats, Lice and History (Little, Brown and Company, Boston, 1935).
14. G. N. Wolcott, personal communication.
15. U.S. Department of Health, Education and Welfare, Report of the Secretary's Commission on Pesticides and their Relationship to Environmental Health (Government Printing Office, Washington, D.C., 1969).
16. U.S. Department of Agriculture, Agricultural Statistics 1976 (Government Printing Office, Washington, D.C. 1976).
17. President's Science Advisory Committee, Restoring the Quality of our Environment (Government Printing Office, Washington, D.C., 1965).
18. E. Hindin, D. S. May, G. H. Dunstan, in Organic Pesticides in the Environment (American Chemical Society, 1966), pp. 132-145.
19. G. W. Ware, W. P. Cahill, P. D. Gerhardt, J. M. Witt, J. Econ. Entomol. 63, 1982 (1970).
20. W. E. Buroyne, N. B. Akesson, Agr. Aviation 13, 12 (1971).
21. N. B. Akesson, S. E. Wilce, W. E. Yates, Paper 71-662, Annual Meeting Amer. Soc. Agr. Eng., Chicago, Ill. (7-10 December 1971).

22. W. E. Yates, N. B. Akesson, in Pesticide Formulations (Marcel Dekker, New York, 1973), ch. 7.
23. Farmer's Use of Pesticides in 1971.....Expenditures, Economic Research Service, Agricultural Economic Report No. 296 (Department of Agriculture, Washington, D.C., 1975).
24. D. Pimentel, J. Econ. Entomol. 54, 108 (1961).
25. D. Pimentel, Ann. Entomol. Soc. Amer. 54, 323 (1961).
26. C. R. Malone, Ecology 53, 507 (1972).
27. L. Adams, M. G. Hanavan, N. W. Hosley, D. W. Johnston, J. Wildl. Manage. 13, 245 (1949).
28. J. O. Keith, R. M. Hanse, A. L. Ward, J. Wildl. Manage. 23, 137 (1959).
29. R. D. Porter and S. N. Wiemeyer, Science 165, 199 (1969).
30. S. N. Wiemeyer and R. D. Porter, Nature 227, 737 (1970).
31. G. E. Burdick, E. J. Harris, H. J. Dean, T. M. Walker, J. Skea, D. Colby, Trans. Amer. Fish. Soc. 93, 127 (1964).
32. R. L. DeLong, W. G. Gilmartin, J. G. Simpson, Science 181, 1168 (1973).
33. W. Helle, Advan. Acarol. 2, 71 (1965).
34. R. van den Bosch and P. S. Messenger, Biological Control (Intext Educational Publishers, N. Y., 1973).
35. P. DeBach, Calif. Citrograph 32, 406 (1947).
36. P. H. De Bach, Biological Control by Natural Enemies (Cambridge Univ. Press, London, 1974).
37. Restricting the Use of Phenoxy Herbicides, Agricultural Economic Report No. 194 (Department of Agriculture, Washington, D.C., 1970).
38. I. N. Oka and D. Pimentel, Science 193, 239 (1976).
39. I. N. Oka and D. Pimentel, Environ. Entomol. 3, 911 (1974).
40. J. W. Ingram, E. K. Bynum, L. J. Charpentier, J. Econ. Entomol. 40, 745 (1947).
41. W. B. Fox, Sci. Agric. 28, 423 (1948).
42. J. B. Rowell, R. I. Agric. Exp. Sta. Bull. No. 320 (1953).
43. R. C. Maxwell and R. F. Harwood, Ann. Entomol. Soc. Am. 53, 199 (1960).
44. S. Ishii and C. Hirano, Entomol. Exp. Appl. 6, 257 (1963).
45. C. Hirano, Bull. Natl. Inst. Agric. Sci. Ser. C 17, 146 (1964).
46. D. R. MacKenzie, H. Cole, D. C. Ercegovitch, Phytopathology 58, 1058 (1964).

47. T. J. Simons, A. F. Ross, Phytopathology 55, 1076 (1965).
48. J. A. Pinckard and L. C. Standifer, Plant Dis. Rep. 50, 172 (1966).
49. L. H. Purdy, Plant Dis. Rep. 51, 94 (1967).
50. J. B. Adams and M. E. Drew, Can. J. Zool. 47, 423 (1969).
51. P. C. Cheo, Phytopathology 61, 869 (1971).
52. Losses in Agriculture, Agricultural Research Service 20-1 (Department of Agriculture, Washington, D.C., 1954).
53. D. Pimentel, J. N. Y. Entomol. Soc. 81, 13 (1973).
54. The Pesticide Review 1970, Agr. Stab. and Cons. Serv. (Department of Agriculture, Washington, D.C., 1971).

9

Post Harvest Losses: A Priority of the U.N. University

Max Milner, Nevin S. Scrimshaw, and H. A. B. Parpia

The U.N. University World Hunger Programme

There is universal agreement that the most serious problem facing the world today is the increasing pressure of world population on the food supply, particularly in the developing countries. Thus, when the U.N. General Assembly in 1973 created a United Nations University whose primary concern was to be "the pressing global problems of human survival, development and welfare," it was almost predictable that when the exercise of identifying priorities for this work was initiated, the first to be agreed to was a World Hunger Programme. This priority program identified for itself the following areas of concentration:

1. Human nutritional needs and their fulfilment in practice.
2. Post-harvest food conservation and technology.
3. Nutrition and food objectives in national development planning.
4. Agricultural production/food and nutrition interfaces.

The World Hunger Programme was established on the recommendation of a working meeting of experts held in Tokyo in September 1975 and an advisory task force for programme planning held in New York shortly thereafter. Work began immediately on identifying competent institutions with which the University should be associated.

At its Sixth Session in Caracas in January 1976, the University's Council authorized negotiations to begin with the first two prospective Associated Institutions - the Institute of Nutrition of Central America and Panama (INCAP) in Guatemala City, and the Central Food Technological Research Institute (CFTRI) in Mysore, India. Both these institutes have twenty-five years' experience of practical service to their regions, with extensive experience in international training and applied research. INCAP has been a leader in the identification of human nutritional needs and in the availability of nutrients from diets. It has also done some work on certain aspects of post-harvest food conservation and on food and nutrition policy planning. CFTRI has pioneered in various areas of food technology within India and as an Asian regional training centre. It has the capacity for designing conservation and processing industries through development, transfer and use of appropriate technologies, particularly those of small scale appropriate for village operations.

At the Seventh Session of the University's Council in June 1976, negotiations with the Nutrition Centre of the Philippines (NCP) were approved. The NCP programme of applied nutrition is putting into practice the lessons from similar but less extensive programmes in many developing countries over the past twenty years.

In approving this initial group of 3 institutions, to which others will be added in due course, the University was guided by the following criteria:

1. They must have demonstrated capability for advanced multi- and interdisciplinary training and research.
2. They must have demonstrated special interest in the capacity for training individuals to conduct mission-oriented, applied research.
3. They should have programmes and capability for practical training of Fellows, sponsored by the United Nations University and selected from various countries, in one or more of the priority areas.
4. They should have adequately balanced practical and theoretical approaches to training and research and be familiar with specific problems

and needs in the designated programme area.
5. They should have, or be capable of developing, linkages with fieldwork programmes in which Fellows may participate in the applied or practical aspects of their formal training.
6. They should be capable of organizing regional and international workshops and conferences.
7. They should be so geographically distributed as to avoid undue concentration of effort and be in a position to respond to specifically identified regional problems.

A central tenet of the University is that support is needed for institutions in the developing world if they are to be of maximum value to their own societies and capable of combatting the brain drain to the industrialized countries. Accordingly, the University places emphasis on the need to organize linkages between institutions that will extend the influence of those centres of research and training in developing countries that have already attained a multidisciplinary competence to solve practical problems. The institutions that have become part of the U.N.U. World Hunger Programme network are precisely this sort.

Identifying and building a network of such centers of competence in the developing world and initiation of a U.N. fellowship and training programs for third world nations in these institutions, is therefore the first priority of this new program. In addition, a start has been made on a series of workshops at several of the International Agricultural Research Institutes, to strengthen the interfaces between agriculture, food and nutrition. The first of these workshops took place in December 1976 at the International Institute of Tropical Agriculture (IITA) in Ibadan, Nigeria, and the second at the International Rice Research Institute (IRRI) in Los Banos, the Philippines, early in 1977 (1).

The purpose of the workshops is to establish a multi-disciplinary dialog among agriculturists and nutritionists that will lead to a better understanding of the nature and significance of nutritional and food considerations in agricultural sector planning. The participants in each are food and nutrition experts from the countries of the region, and persons concerned with agricultural sector policies, agricultural extension, and plant breeding. Emphasis is also being placed on post-harvest food conservation, including more efficient storage and processing of food at home and village levels.

Why Post Harvest Losses Must Be Dealt With

The programme recognizes that increased food production will not on its own solve the problems of hunger and malnutrition, however. There is a problem of more equitable distribution, since the world already produces more than double the quantity of edible materials necessary to fulfil human nutritional needs, and yet as many as half the world's peoples have inadequate supplies of some or all foods.

The industrial and petroleum exporting countries either produce enough food for their needs or have the purchasing power to obtain it from countries with marketable food surpluses. Conversely, most developing countries are falling behind in *per capita* food production and cannot afford to import the additional food they require.

Low-income groups in many areas simply have so few resources that they cannot produce or buy enough of even the simplest foods to satisfy their nutritional needs. The situation is so precarious that in times of crop failure, drought, or unemployment, famine must inexorably follow. Clearly, such low-income groups can be helped only by increasing their income or by providing them greater access to the means of food production. No early solution to the prevalence of poverty is in sight because of the complexity of contributing social, economic, and political factors, which are resistant to change.

On the related questions of food production and population growth, there is already extensive international and bilateral activity, and international agricultural agencies are already directing far more resources to meet these problems of food production than will be available to the United Nations University.

Very little international effort, however, is applied to the question of what happens to food after it has been produced. Most research and applied science relating primarily to agricultural production stops at the farm gate. The qualitative and quantitative losses of between 20 and 40 per cent of the food which occurs in developing countries have received little attention.

To start with, there are large physical losses of food during storage, handling, and transportation to rodents, insects, mould, and simple spoilage. Again, estimates vary considerably since reports, generally of anecdotal character, are often incomplete or unreliable, and averages conceal national and annual extremes. It has been reported, for example, that in the year 1953/4 on the island of Mindanao in the Philippines rats were estimated to have consumed 90 per cent of the rice, 50 per cent of the maize, and 50 per cent of the cane sugar. In India, one graphic summary suggested that it would take a train 3,000 miles long to haul the grain eaten by rats in a single year. The judgement of individuals who monitor this problem is summarized in Tables 1, 2, and 3 from papers by Parpia (2) and Spurgeon (3). Clearly, however, much effort is needed to establish the true nature and scope of such losses.

Post-harvest food and agricultural technology is an interdisciplinary science whose functions begin after crops are harvested, animals slaughtered, or fish caught. It covers handling, storage, processing, packaging, and transport, as well as the distribution and ultimate use of food. Its main objective is to contribute towards solving world food problems through interdisciplinary application of science and technology and management practices in order to conserve food and improve, to the maximum extent possible, its nutritional quality to meet human needs.

The Need for Research

Although it is recognized that food losses due to pests, spoilage, and wasteful processing procedures are high, especially in developing countries, and that their prevention deserves urgent attention, the information available on the nature and magnitude of food losses is limited. Collection of more data on this problem in selected countries would help to clarify the role of post-harvest food losses in producing food shortages and increasing

Table 1. Some Estimates of Losses in Different Countries

Country	Material	Loss: Percentage[a]	Loss: Value
Nigeria	Sorghum	46	
	Cow pea	41	
United States	Stored grain		$ 500 million
	Packed food		150 million
	All crops		3500 million
India	All grains		
	Field loss	25	
	Storage loss	15	
	Handling & processing loss	7	
	Other losses	3	
Germany	Harvested grain		DM 71.4 million
Sierra Leone	Rice	41	
	Maize	14	
Tropical Africa	All crops (storage and handling)	30	

[a]These percentages refer to postharvest losses unless otherwise stated. Although the figures refer to specific crops in most cases, they are sufficiently indicative to lay emphasis on the problem of food losses (from Parpia, 2).

Table 2

Estimated range of losses from a variety of causes in the postharvest system of a number of countries during storage of various crops.[a]

Crop	Country	Weight loss (%)	Period of storage (months)
Legumes	Upper Volta	50-100	12
	Tanzania	50	12
	Ghana	9.3	12
Maize	Zambia	90-100	12
	Benin	30-50	5
	USA	0.5	12
Rice	Malaysia	17	8-9
	Japan	5	12
	United Arab Republic	0.5	12
Sorghum			
(unthreshed)	Nigeria	2-62	14
(threshed)	USA	3.4	12
Wheat	Nigeria	34	24
	India	8.3	12
	USA	3.0	12

[a]After Hall, D.W. Handling and storage of food grains in tropical and subtropical areas. Rome, Food and Agriculture Organization of the United Nations, FAO Agricultural Development Paper No. 90, 1970, 20-21 (from Spurgeon, 3)

Table 3
Estimates of quantitative losses during handling and processing of rice in Southeast Asia.[a]

Operation	Range of losses (%)
Harvesting	1-3
Handling	2-7
Threshing	2-6
Drying	1-5
Storing	2-6
Milling	2-10
Total	10-37

[a]D.B. de Padua, University of the Philippines at Los Banos, College, Laguna, Philippines, personal communication, 1975 (from Spurgeon, 3).

prices and would identify the economic and social benefits such countries would gain from the reduction of such losses. Work in this area would be of great value in helping to decrease food losses at a time when production costs have increased as a result of a nearly threefold rise in the cost of energy, fertilizers and other inputs. The study would help policy makers give high priority in development plans to the prevention of food losses and to allocation of adequate resources for the solution of this problem at national and international levels.

Many traditional technologies already exist in developing countries for the conservation and processing of food, but means must be explored to increase their impact. In considering the post-harvest conservation of food, it is desirable to survey, identify, and understand these technologies with a view of transforming them into modern, science-based ones. Such technologies would not only be socially appropriate, but also may have greater user acceptance, particularly in rural areas, than many newly developed processes.

The food distribution and marketing system should also be considered in plans for preventing wastage. A large majority of the populations in developing countries is outside the market economy as it is understood today. There is a need to develop a system of marketing or distribution to reach these consumers with the objectives of reducing present food waste, minimizing cost, and providing foods needed to prevent malnutrition.

Because grain resources often are used for producing animal products at quite low conversion yields (20-25%), which many developing countries cannot afford, efforts should be devoted toward providing alternatives or supplements to animal products in the diet by means of vegetable analogues or extenders. Technologies should be further developed to recover, for human consumption, food components of high biological value in commodities used traditionally for animal feeding. Means should also be sought for using resulting non-food by-products for animal feeds. Animal breeding efforts should attempt to develop livestock breeds capable of efficiently utilizing these products.

The present use of oilseeds primarily for animal feeding is an example of this situation. Oilseeds represent a valuable protein resource and their utilization through animal feed is not always economical or justifiable when it has been demonstrated that they can be successfully used in the manufacturing of infant and weaning foods and milk substitutes. Similarly, rice bran, an essentially unutilized byproduct of the rice milling industry, can provide high quality fat to overcome shortages of concentrated dietary energy sources, and the extracted bran used for animal feed.

The consumption of protein of low biological value as is the case in many countries whose primary food staples are grains and tubers results in waste because it cannot be used efficiently by the body. This loss can be minimized by combining foods whose proteins mutually complement each other and/or by providing foods that are individually superior in their nutritive content. In certain cases, this could also be achieved through enrichment or fortification. Research to develop balanced foods and diets would help prevent such waste of protein.

Food and agricultural products processing industries produce fairly large amounts of residues. Improved utilization of these wastes by developing non-conventional foods could enhance food supplies and bring other benefits as well. For example, the disposal of molasses from processing of sugar cane creates significant pollution problems which can be prevented through better utilization of this byproduct as feed and food.

As a long-term measure, particular attention should be given to the use of non-food carbohydrates like molasses as well as to cellulosic residues and hydrocarbons, all unsuitable for human consumption, as energy sources for the manufacture of single-cell protein products for use in animal feeds. Newer technology will permit the processing of these proteins for direct human food use. Such developments are also relevant in the U.S. where we can already clearly identify growing constraints on our land, energy and water resources.

Growing Concern for the Problem

Research and other actions being considered by the UNU comprise only a part of any adequate attack on this global problem. FAO has been conducting programs in this area for a number of years and is now giving priority to prevention of post-harvest losses (4). The Tropical Products Institute in the U.K. has been a productive research and project implementation group, particularly in the tropics. In Africa 6 agencies are providing help to countries through the Group for Assistance on Storage of Grains in Africa (GASGA). More recently the International Development Research Center in Canada has sponsored research and corrective programs in the post-harvest losses and processing areas. Five Southeast Asian countries have begun a joint program through the Southeast Asia Regional Center for Agriculture (SEARCA) aimed at devising new or improved post-harvest systems. Kansas State University, under the auspices of AID, has been leading such activities in a number of developing countries. Recently AID has asked the National Academy of Science to conduct a survey of the scope of such losses and an evaluation of possible interventions. On behalf of AID, the American Association of Cereal Chemists is engaged in a study to develop assessment methodology for grain losses. It seems clear that there is now a growing momentum in this important area so vital to increasing world food supply.

References

1. Interfaces Between Agriculture, Nutrition and Food Science. United Nations University, World Hunger Programme, Monograph Series, (forthcoming) 1977.

2. H.A.B. Parpia, Postharvest losses - impact of their prevention on food supplies, nutrition and development. Chapter 18 in Nutrition and Agricultural Development - Significance and Potential for the Tropics, pp.195-206, Ed. by N.S. Scrimshaw and M. Behar. Plenum Press, New York and London, 1976.

3. David Spurgeon, Hidden Harvest: A systems approach to postharvest technology, IDRC-062e. International Development Research Centre, Ottawa, Canada, 1976.

4. Food and Agricultural Organization of the United Nations. Reducing Postharvest Losses. COAG/77/6, February 1977. Item 6 of the Provisional Agenda, Committee on Agriculture, Rome, 20-28 April, 1977.

Index

For Product Safety Concerns and Information please contact our EU representative GPSR@taylorandfrancis.com Taylor & Francis Verlag GmbH, Kaufingerstraße 24, 80331 München, Germany

Batch number: 10725937

Printed by Printforce, the Netherlands